Tools of Systems Thinkers

Learn Advanced Deduction, Decision-Making, and Problem-Solving Skills with Mental Models and System Maps.

Written by
Albert Rutherford

Gift Alert

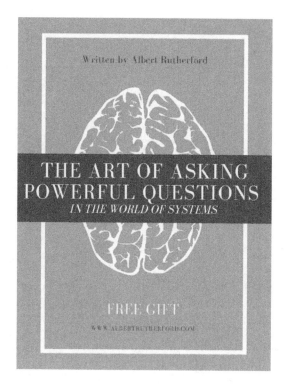

Visit www.albertrutherford.com and download your FREE copy!

Table of Contents

10

Endnotes

Introduction

Albert Einstein once posed this question: "Have you ever thought about how you think?" A lot of people would say "no." After all, when was the last time we stopped to analyze our own thinking? Herein lies the problem. Einstein pointed out that certain issues cannot be solved with the same kind of thinking that created those issues in the first place.

Certainly, second- and third-guessing every thought crossing our mind is not a reasonable expectation. It is time-consuming and, frankly, tiring. What we can do, however, is learn and memorize—so to say—new thought patterns and use them to our benefit. My mission in this book is to make a shift in your cognitive assessment style to develop the most optimal solution to the problems in your life. The rewired thinking path I'm talking about is called systems thinking.

What Is Systems Thinking?

People tend to view the world as an accumulation of separate, independent elements. This is the grocery store, those are the products in it. We like to buy shallots instead of onions. There are more gas stations in our area than in the city nearby. We take these elements of life for granted without much questioning. But why are there more gas stations in our town? How do the shallots get to the supermarket? How are these things connected?

Systems thinking is a way of viewing the world as a cluster of interdependent systems. Think of it as a large machine in which one cogwheel drives the next. Everything is related to each other, no matter how far-fetched it may seem. The big supermarket we have in our town attracts a lot of buyers from the surrounding smaller ones. Many suppliers come and go daily to restock the shelves of the supermarket with all sorts of goods. For this purpose, gas stations tend to cluster in this area. There is high traffic—literally and figuratively.

But this explanation may not offer a deep insight into what systems thinking is. To clear things up, let's discuss what systems thinking is not.

Our brain has created an abundance of mental shortcuts throughout our evolution. Cavemen used these shortcuts to help them make sense of their world. Quick decisions have saved their lives many times. Over thousands of years, our brains were conditioned to make decisions quickly. The frequent alternative was dying a sudden and violent death.

Evolutionarily, therefore, our way of thinking developed to be linear and focused. This is good because it does not take much mental energy to understand the world around us. Our brain knows how to split external stimuli into bite-sized chunks.

You see, our brain uses roughly 20 percent of our energy. It makes sense that we need to be as efficient with our mental usage as possible. But this way of thinking is restrictive and may not be sufficient when we deal with more complex problems in our lives. Most of the time, we only treat the symptoms instead of addressing the root cause itself.

Here's an example. Suppose that you have problems sleeping at night. That is the symptom. Treating the symptom would mean something like taking sleeping pills and hoping for the best. But this is only a temporary solution. The cause of your sleeping trouble might be that you have too much coffee in the afternoon, or that you are feeling anxious about your job.

If you struck at the heart of the problem, you'd be able to solve it for good. You could stop drinking coffee after lunch or address your work-related issues head-on. But if you only tend to the symptoms, it might lead to more problems. For example, using sleeping pills may lead to addiction, or your work-life issues may escalate. Systems thinking challenges the mainstream, reductionist view. It introduces an expansionist idea, which means seeing the world as a massive network with all its interconnected elements. Every little aspect of it is important.

By understanding how each element relates to another, we can comprehend the larger system at play. We can then start to understand how it works and identify opportunities for solving our problems.

The Three Key Systems

We can distinguish three key systems at play: social, industrial, and ecosystem. All three work in conjunction to keep our civilization going. Without these three, society would devolve into chaos.

Social systems comprise all the rules, norms, and structures that we, humans, create. They govern how we interact.

Industrial systems consist of all the things that sustain our lives, including food production and manufacturing of products, which require the extraction and processing of natural resources.

Finally, there is the ecosystem. This provides the natural resources we need to thrive, such as air, water, food, and other minerals. In other words, the ecosystem supplies and sustains the other two critical systems. As you can see, life as we know it would not work without these three systems' fragile balance.

The systems thinking approach uses the big-picture view on a large scale. It considers how changing one element in any of the systems mentioned above can influence another. It is no wonder why many people feel intimidated by this way of thinking or are simply overwhelmed. The payoff is that once you make sense of the whole picture, it is easier to find out the root problem and resolve it once and for all.

Why Has the Linear Thinking Approach Been So Dominant?

Linear thinking is characterized as "A leads to B, which results in C." This seems to be a logical way of thinking, and we love it because it makes the world look very simple. In fact, we have created many of our regulations based on this way of thinking, from the scientific method (cause and effect), to rigid directives from insurance companies, and to governments' strict laws.

These linear methods are designed to keep order. But the problem with this approach is that the complex challenges of life don't exist in isolation; they are interconnected.

For example, someone has regular panic attacks (A). They visit the doctor (B), and the doctor gives them a medication to solve the problem (C). This is a linear process. But often, a pill doesn't fix the root cause of the problem. Maybe this person has mental health issues because they face difficulties at work. Their marriage is falling apart, and they can't sleep … These reasons might overwhelm the person to the point that the person can't cope with them and then has a panic attack. Doctors tend to isolate the health problem. If one has a panic attack or severe anxiety, they put a Band-Aid on the distressing emotions, solving it with a pill. But a Xanax won't fix the root cause of panic attacks and anxiety. The patient might be at the clinic again in a few days or weeks with another panic attack. The immediate problem, the symptom, will need treatment again.

This way of thinking isn't limited to the health care system. Social institutions are often grouped in different departments that hardly communicate with each other. Problems such as the one described above are present across various areas from education to economy to sustainability, leading to a wide variety of challenges. For instance, the education system has distinct specialties (biology, math, psychology), which are separated to focus on one particular scientific area. Companies are usually focused on one exact issue in the commercial sector that they try to solve through their specialized products.

The isolation of elements often leads to skewed prioritization. For example, you want to make the world a better place for all. Where should you start? An economist will argue to begin with the financial sector, a sociologist will urge to start with human behavior, while an environmentalist will create a case for conserving nature. Personal perspectives will lead to competition and conflicting strategies. Everyone sticks to their own view hoping that the other parties will give in. But unfortunately, this attitude leads to parallel problem-solving attempts often in conflict with each other.

Linear thinking is enforced by our education system. We learn to reduce and dissect problems to smaller, more manageable components without looking at the big picture first. Our learning material is also distributed across relatively isolated departments and sciences. We are never really taught how math and biology interrelate. Why are they both important to develop a well-rounded picture of the world?

However, suppose we shift to a systems thinking lens. In that case, we will see that everything is interconnected and dynamically changing. We need to integrate the separate departments of knowledge into one to see how they are connected. This way, we can better understand the complex challenges of the world.

Systems thinking approaches problems integrally and holistically. It focuses on the whole system, encourages communication between its elements, and supports joint problem-solving based on real priorities.

This might sound difficult, but luckily, humans have an intuitive understanding of the world's complex, dynamic, and interconnected systems. Namely, we live in this world every day. We just need to use adjusted thinking lenses to shift from the linear, one-dimensional and into the nonlinear three-dimensional reality.

Today the urgency to solve complex, chaotic, and systemic challenges is paramount. These problems cannot be fixed with a reductionist approach. The issues we face today are complex. In problems such as racism, homelessness, and global politics, many elements play a role. How can we approach such challenges effectively?

System Boundaries

When everything is interconnected, where should one draw the boundaries of a problem? It can pose difficulties to know where one problem ends and another starts. Our intervention to fix the issue is much more thorough and targeted if we define well what we're trying to solve. For instance, our friend's problems with panic attacks might be rooted in an unhealthy marriage. That's the area that needs solutions. If that bottleneck issue is resolved, his work- and sleep-related problems will fade, as well. But if we don't dig deep to define his marital problems' boundaries, all of it can seem like a mental wormhole.

We need to properly investigate and assess what's happening and then apply a scope to the problem. When it comes to finding interconnections, the possibilities can go on and on. That is why it is important to find limits. In other words, a system boundary is a line dividing the concerned system from everything else. By investigating which parts are relevantly interacting with each other, we can understand how a problem loops around in the system. Then we can identify the critical elements involved. Inside the scope will be all the elements that we need to consider. Outside of the scope will be the other elements that are not relevant in this case.

A university's system boundary can be defined by the campus it owns, its student body, and the staff working therein. This way, we create a direct boundary; we name the system (the university). We can also draw a boundary discerning what is not the part of this system, say, the other university across the street. However, if we expand our assessment field from the university campus to the whole education system, then the universities on each side of the road would be a part of the system.

It is a matter of perspective and the situation we are dealing with, which helps us identify what is part of our system and what isn't.

The System as a Whole

I'm about to present an example borrowed from Draper Kauffman's introduction to systems thinking from the 1980s.

Imagine that you have a cup of milk. How can you have more milk? You could buy more from the supermarket and add it to your existing amount. Thus you'll get more milk. However, if you have a milk-producing cow and buy another cow, you won't end up with a larger cow. You'll have two cows. And two cows give you more milk. Now, let's go in the other direction.

Say you need less milk, so you pour half of what you have into another cup. If you chop your cow in half, though, you won't get two smaller cows. You'll get a fair amount of beef and some horns for decorative purposes. But the cow as a system (the cow is a system, too) will cease to exist in its previous form, as it was drastically altered.

Systems function as a whole. They lose their function if you cut them in two. And modifications on a subsystem level can lead to these drastic, crippling changes. Take the ecosystem as a whole; if you remove the grass from it, the cow won't eat, and it won't produce milk. This is an oversimplification, of course, but you can see now how everything's so vitally interconnected.

Thus, if we want to interfere and make changes mindfully in a system, we must understand its interconnections—its flows. Isolating one element won't make a big difference and won't solve the problem. Cutting the proverbial cow in half may destroy the entire system. We need to find a way to intervene somewhere between these two extremes—to guide the flow of dominant dynamics.

Five Key Concepts

Systems thinking requires the use of specific terms to describe given processes. Just like legal terms, some of the words mentioned here might seem nonsensical. Understanding them, however, is essential to be an effective systems thinker.

1. Interconnectedness

Systems thinking needs us to change our mindset, from linear (A to B) to circular (A to B to C to A), to adequately represent interconnectedness. This circular mindset allows us to make sense of the complexity of our world.

Think of it like the food chain, in which everything relies on something else. We need food, air, and water to survive. Food comes from animals, which consume plants. When animals or humans die, they become natural fertilizers for plants. And there goes the circle of life, just like Mufasa taught us. We can start to look at the world differently.

2. Synthesis

Synthesis refers to the creation of something new by combining two or more existing parts. Synthesis is the opposite of analysis, which seeks to dissect complex questions or problems into small, manageable chunks.

As mentioned before, all systems in life are interconnected. As such, a more effective solution to any problem would be one that is holistic in nature. Synthesis allows us to see and understand the big picture, its parts, and how they interact. Synthesis is the goal of systems thinking.

3. Emergence

A larger whole is comprised of many smaller parts. Emergence describes the phenomenon when these smaller parts unite to create the whole. Emergence, therefore, is the product of synergy. You can say that emergence shows how everything in life results from the interaction between two or more elements.

A good example here is the Big Bang. As we know, life is a product of self-organizing forces that brought countless atoms together to form objects as small as a pebble, to something vast such as galaxies.

Understanding the importance of emergence, you might start to connect the dot between two seemingly unrelated elements and arrive at conclusions you wouldn't have seen with linear thinking.

4. Feedback Loops

Because everything is interconnected, the whole system is comprised of a series of feedback loops. When we understand the nature of feedback loops, we can start to plan interventions to resolve the problems they create.

There are two types of feedback loops in systems: reinforcing and balancing.

In a reinforcing feedback loop, the elements strengthen or reinforce the same behavior. Take weight gain, for example. If we don't exercise, we will gain weight, which will make it harder to move around. This discourages us from exercising, which leads to more weight gain.

It also works in reverse. If you exercise, you will lose weight, making it easier to move around. The success will encourage you to exercise, leading to more weight loss.

As the name suggests, in a balancing feedback loop, each element works to balance out the loop's behavior.

Mother Nature finely displays this dynamic, maintaining the balance between predator and prey populations. The lion population regulates the deer population by hunting them. The deer population controls the lion population in return because they are the lions' food source. If the number of lions rises too quickly, some will starve and die, which brings the number of lions back down. If the number of deer increases too fast, the number of lions will also rise, which means more deer are killed. This brings the deer population back down, which will affect the lion population and so on.

Suppose you take away too much of any of the lion or deer population. Your actions will cause a reinforcing feedback loop that can lead to a population boom or rapid decline.

5. Causality

Feedback loops give you insight into causality, which explains how one thing can lead to another. Everything we do has consequences—cause and effect. In systems thinking, causality is about understanding how one element influences another in a system. Being able to track causality is invaluable when you try to map out systems.

Chapter 1: The System Mindset

One of the most significant leaps in understanding and solving both the intricate and everyday problems of life came along with the development of systems thinking. MIT professor Jay Forrester realized the need for a better perspective on worldwide problems. In 1956 he started outlining a new approach that today we call systems thinking.[i] Just like engineering principles help us better understand mechanical structures, systems thinking helps us gain deeper insight into social, economic, and ecological systems.

Forrester's approach is fundamentally different from that of traditional analysis. While the latter focuses on dissecting and studying individual and separate parts, systems thinking encourages us to notice the relations and connections between every element. It can be said that systems thinking is like seeing the "big picture," instead of only a part of it.

Let's see an example of a systems thinking cognitive perspective: reducing crop damage caused by insects. When an insect is damaging our crops, the traditional response is to use pesticides to eliminate them. Assume there is a fully effective pesticide that can kill all the insects and has no side effects on humans. Would it help the farmers in the long run? Who can tell? Based on historical data, the pesticide solution's problem is that it has decreasing effectiveness. As pests develop immunity against it, we need to use more and more pesticides. Or, it can also facilitate the spread of another insect type that is left without a natural predator, given the initial insect population was eradicated. Pesticides can make the whole situation even worse than before in the long term.

Looking at problems through a systems thinking lens can help us foresee, prevent, or handle unintended consequences by developing a more holistic view of the problem at hand.

To understand what systems thinking is, we need to know what a system is. The dictionary defines a system as such: "A regularly interacting or interdependent group of items forming a unified whole."[ii] A system collects different elements that have relationships with each other. These elements are affected by the interactions happening within the system. For example, a company is a system. The elements in this system are the employees, managers, the CEO, but also the customers and competitors. The term "unified whole" refers to the system having a fusing purpose above the interactions. These elements interact and affect each other for and with a purpose. In the company's case, this purpose can be maximizing profit, usefulness, product quality, etc.

If you think about it, most of the real world's structures and arrangements are suffixed by the word system: mechanical system, education system, and respiratory system. This tells us that systems can be found anywhere and everywhere, from our own body to an inanimate object. Furthermore, everything in this universe is related and connected in some way.

A system needs to have elements, interconnections, and boundaries. Emergence— when certain parts interact, forming new connections—is frequently present in systems.

Another attributive of systems is their constantly changing, dynamic nature.

A system must stay whole to fulfill its function or purpose. Remember the system of the cow?

The cow is a system because it fulfills all the requirements to be considered one. For boundary, it has its skin, and interrelationship can be seen between the organs' interactions. The cow is also part of a more extensive system, such as the food industry. The food industry is part of a bigger system, namely the economy. All these are held together by the ecosystem. The animal consumes grass and water. The food industry relies on the economy, which needs the ecosystem in many ways.

Systems are also identified by their inability to exist with one or more of its parts gone. Think about an airplane's engine. If you take the engine off, the airplane wouldn't be functional. Hence it would cease being a system with the function of air transportation. You can deduce similar conclusions if your computer or television don't get electricity. They wouldn't be able to fulfill their role.

Systems Thinking Analogies

Let's take a look at an example that can help you deeper understand and learn about systems thinking.

The analogy I want you to think about is the connection between walking and climbing Mount Everest. When you are a toddler, you are just learning to walk in your own home. Surrounded by loved ones who protect you and encourage you to explore more, you may start fantasizing about grandiose plans. But you don't have the skills and strength to do them—yet. If you want to climb Everest, walking is the first skill you have to master. As you grow, you find walking more effortless, and you can balance yourself perfectly. Walking becomes a subconscious activity at this point.

To speed up the learning process, you can attend some climbing events. This way, you build strong skills to climb Everest one day.

The message is that we always start from smaller goals and gradually work towards bigger ones. This is how you develop your systems thinking skills, too. You learn the basic concepts to build mental models and design solutions that turn economies around.

My second analogy is how systems thinking relates to a telescope and a microscope. Just as we can see the vast landscape through the telescope, we can see the big picture through systems thinking. We can also zoom in and notice the small changes at sub-systemic levels the same way we would use a microscope.

A third analogy is known as the lensing tool. This is a more in-depth approach to observe systems by changing our perspective, just like an animator who can show a wide range of emotions with only a few strokes. They can make a character go from happy to sad by shifting facial expressions. Lensing involves transforming your current perspective into a different one to gain more insight into the situation at hand.

Systems Thinking Terms

In this section, we'll learn the nitty-gritty of the systems thinking language. I will present the discipline's fundamental terms. If you are already familiar with this information (if this is not the first systems thinking book you read), you may skip this part. But you can just as well stick with me and refresh your knowledge on key concepts like feedback loops, systems dynamics, system delays, and more.

My goal here is to help you master the terminology used in systems thinking.

The purpose of learning these concepts is to develop the ability to communicate with fellow systems thinkers about the complex issues you're examining. By understanding the structure, you can develop high-leverage intervention strategies to address the problem's root cause. By learning systems thinking language, you'll be able to verbalize your predictions and map out structures that can be a catalyst for problem-solving.

As we can see, systems thinking can help us

- solve existing problems;
- find improvements;
- and prevent future problems.

What is Systems Dynamics?[iii]

Systems dynamics show the nonlinear behavior tendencies of complex systems. With the help of stocks, flows, feedback, and delays, we can detect how quickly changes happen within the system. Often, the behavior of the whole system can't be explained based only on its individual parts' behavior. Determining what kind of interactions are at play between the elements is very important. Stocks and flows are the building blocks of the systems dynamics model. We will learn a lot more about them in the upcoming chapters.

Variables or Elements

Variables or elements are usually neutral nouns presenting situations where behavior patterns change over time.

David Peter Stroh, a systems thinking expert, defines variables as "the nouns of systems thinking language." They are the fundamental building blocks of a systems story, impacting each other, and changing over time. Variables (or elements) are critical in building a clear understanding of what has been happening and why. Systems thinking has the advantage of being able to combine quantitative variables such as profits with qualitative variables such as morale.[iv]

Variables in systems thinking language have a few strict characteristics. Besides being neutral nouns, they are not events. The value of a variable can go up and down over time. The nouns you use for variables should incorporate the level of change over time. Thus, using noun phrases like "the level of ..." or "the quantity of ..." is indicated. Avoid using verbs of action and nouns where the level of change is hard to measure (for example, nouns like culture, religion, etc.) as variables.

When we talk about neutral nouns, we want to avoid using evaluation words such as rich, bad, or good. The suggestion of an amount should also be avoided when defining variables. Words like increasing, lack of, and abundance of should not be added to the variable.

Links

Links show the causal connection between two variables. The change of one variable causes a shift in the other.

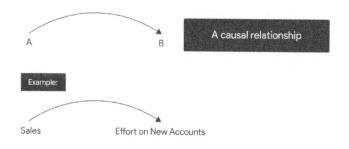

Picture 1: How to illustrate links.

To discover and understand links, one should select a variable and ask:

• What are the possible causes for the increase or decrease in this variable? Name these factors as variables (nouns, something that can increase or decrease).

• What are the possible consequences or effects of an increase or decrease in this variable? Name these effects as variables.[v]

Usually, there is more than one cause or effect when we answer these questions. We should focus on those causes and effects that best explain our field of interest. In some cases, the change in one variable could cause an increase and a decrease at the same time in another variable. For example, when the government wants to raise taxes to increase its budget, two things will happen: On one hand, the governmental budget will increase due to the higher tax inflow, but on the other hand, the government will also experience pressure from citizens who, for their higher tax payment, will demand better government facilities. If the government addresses these pressures, the governmental budget will experience a decrease.

Let's take a look at some simple links:

Our level of thirst determines the amount of water we drink. As the level of our thirst changes, the amount of water we drink changes. In other words, the variable of thirst is linked to another variable, water intake. The link between these two is the process of drinking. The more we drink, the less thirsty we'll feel, and thus the less water we crave.

Level of Thirst → Amount of Water Intake

Direction of Links

The links are the little arrows that connect the elements of the causal loop diagram. We can label these links as positive or negative, using + or − to mark them. Or we can name them links pointing to the "same direction" or "opposite direction," marking them *s* or *o*.

Both marking methods are correct. It's a question of preference and need which one you choose. Let's get a bit more familiar with them.

s/+ and o/−

s/+ represents an action that generates a result flowing in one direction, which then creates a consequence flowing in the same direction. For example, the more hours a student invests in studying, the better they will perform at the test.

$o/-$ shows that an action leads to a consequence in the opposite direction. For example, the more hours a student invests in studying, the less time they will have to socialize.

In traditional system dynamics modeling, the polarity of causal links has been notated with + and −. Later, in systems dynamics development, people started using the s and o demarcations. Some people argue that + and − are a better way of labeling links because they can be used both for regular causal links and stock-and-flow links alike.

I prefer to use s and o links wherever I can in my books for beginners for two reasons:

1. They prevent the mislabeling of polarities. Beginners tend to be susceptible to this when they trace the up-and-down implications of a change in an element within the loop. Imagine you get to a link where it makes sense to say, "When A drops, B usually grows." Here, the gut reaction is to mark this link with a positive sign, meaning "growth," instead of a negative sign pointing to a "change in the opposite direction."

There is an easy fix to this tendency, though. If you mark each link's polarity separately and then immediately assess the implication of an increase in the element by the tail of the arrow, then the direction of change in the element at the front of the arrow automatically matches the correct polarity of the link. Say, "If A grows, then B usually tends to drop." In this case, we can see that the link is negative. Once each link has been assigned a polarity, we can assess whether the loop is reinforcing or balancing. (I talk about how to do that in a bit.)

2. Another reason I prefer to use *s* and *o* over + and − in the beginning is because some people tend to attach the meaning of good and bad to + and −. While a more trained systems thinker automatically detaches from such labels, these associations can be influential for a beginner.

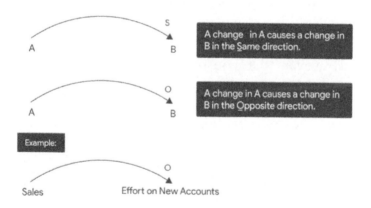

Picture 2: How to illustrate links with directions.

In Picture 2 we can see that as a company's product sales go up, they get less motivated to invest energy in adding new accounts.

If we looked at our level of thirst and the amount of water we drink, they would have a link pointing to the same direction. As the level of our thirst increases, the amount of water we drink increases as well. Similarly, the less thirsty we are, the less we drink.

However, the consumption of water and our thirst can have an opposite link relationship as well. Namely, the more water we drink, the less thirsty we feel. Or, as the amount of water consumed decreases, our level of thirst increases.

When you design a feedback loop or causal diagram, it's good to indicate your links' direction. This practice makes the diagram easier to follow and helps discover possible missing variables that could add necessary information to the analysis.

Delay Analysis

Changes that happen within a system often bring about some delays. When we think of a business, we can distinguish three types of delays: perception, response, and delivery delays. These have a significant influence on the behavior of the system. Delays can make or break the quality of long-term trends. In balancing feedback loops, delays can cause oscillation, which makes it hard to stabilize the system. In reinforcing feedback loops, delays can prevent interventions from taking firm root by creating a circular, reinforcing process.

Unfortunately, analysts often ignore delays and their negative consequences when making a diagnosis about a problem. When the harm caused by delays becomes apparent, people often react swiftly to put out fires. Such hasty interventions are hardly ever proportionate or well-tested, resulting in over- or underperformance.

Let's see how the three types of delays could affect a bookstore.

Rob, the bookstore owner, has to monitor his stock of books, meaning he watches the supply-and-demand trend to make decisions about his inventory. As Rob analyzes this trend's behavior, no matter how wisely he may try to balance the supply and demand, there will inherently be delays in the process.

First, Rob has to take into account the "perception delay." This can be an intentional or unintentional delay. In the case of analyzing inventory, it is often intentional. The owner of a store is trying to decide whether or not to order additional stock. He does not want to immediately react to every small blip of an increase or decrease in sales. Before increasing his inventory, he wants to average the sales for at least a short amount of time to differentiate actual sales trends from just a temporary uptick or downturn.

Next, there will be a "response delay." Once Rob knows which books need to be reordered, he doesn't want to make the complete adjustment in one single order. Being a very cautious business owner, Rob only makes partial adjustments over a short period to ensure his trend assessment is accurate.

Finally, there is a "delivery delay." This delay is out of Rob's control but must be accounted for in his ordering decisions. When Rob places an order, it will take some time for the publishing company to receive, process, print, and deliver the order.[vi]

As the order arrives, Rob will have to continue to carefully monitor the sales trends to check if the previous order was a good bet. If it was not, he needs to optimize his future orders.

Invariably, some mistakes will be made because it is impossible to predict what customers will do in the future with complete certainty. No matter how experienced Rob is, he will need to make adjustments continually. Not because he was careless or ignorant, but because, try as he might, there will always be a small blind spot in information and delivery delays. The best intervention to improve a system's (in our case, the bookstore's) performance is to shorten or get rid of delays.

Please note that delays are illustrated on systems diagrams by drawing a pair of parallel lines (//) on the affected link.

Let's see how to apply systems thinking in practice. To do so, follow these simple steps:

1. Identify and confirm the system in which you want to intervene.

Think about what you are exploring, and figure out its boundaries. For this, you can take a piece of paper and start a cluster map, or you can do it with other preferred systems mapping tools. (We learn more about them in the next two chapters.)

2. Start formulating the problem. Identify the elements, interconnections, as well as the desired outcome following your intervention. This will eventually turn the whole big picture into a multitude of small snapshots.

3. Start asking questions to discover the point of view of others. Explore and go deeper for more insight with each question you ask.

Conclusion

On your journey of becoming a systems thinker, you'll go through a steep learning curve. You need to master some definitions and invest time in the practical application of what you've learned. Slowly you'll be able to identify and demarcate the system—or sub-system—you wish to explore.

Complex problems surround us in our work life, at home, in society. They can be tricky to deal with. The traditional and conventional solutions may aggravate our issues instead of alleviating them. This is where systems thinking comes into play. It raises our awareness to a point where we, as individuals or organizations, can develop better solutions to problems and create lasting changes.

Exercise

Let's do some practice. Look around yourself right now and try to identify systems. For example, your home has several systems. You can see the internet router and the computer. In the kitchen, the microwave and the refrigerator. All of these appliances run on electricity. The overarching system holding these "sub-systems" together is electricity. But your home itself is a system. Or the street your home is a part of. Or the town your street is a part of, and so on.

Imagine a more extensive system now—the population of your country. Within this system, you can explore the changes that are happening. More births mean an increase in population and more deaths lead to a decrease therein. You can explore what triggers each event and how you would intervene to stabilize or influence the changes in any direction. Think about China's one-child policy trying to slow down population growth. In contrast, in communist Romania, birth control of any sort was prohibited to boost the population. Both interventions created a long list of negative consequences such as gender imbalance in China (as sons were preferred over daughters) and the orphan crisis in Romania.

Find at least five more systems around you.

Chapter 2: Mental Models and Mind Maps[vii]

Introducing the Idea of Mental Models

We are surrounded by models. We use them to predict elections, forecast economic performance, and even the fashion industry shows how certain clothes would look on a person. Doctors use models to make a diagnosis and structural engineers create the foundation of buildings thanks to them.

And in systems thinking, we use models to illustrate the interconnections inside a system. We can create a model to see how an actor or dynamic will perform over the long run; we call these models behavior-over-time graphs. Cluster maps model a series of seemingly unrelated ideas and find connections between them. We can illustrate behavior in general through stock-and-flow diagrams, and so on. We will learn more about each of these models in the following chapters. But first, let's see ...

... What *is* a model?

If we take the meaning literally, the word model means "a smaller version of something big." Think about the rudimentary war plans our ancestors made. The chief took a wooden stick and modeled the war field into the sand. He explained where forces should be split, concentrated, a bigger rock they could hide behind, and so on. We recognized early the value of simplification. It is much easier to understand and explain things in a visual way.

Maps are models. They are meant to model the geographical environment of a given size of land on a small piece of paper. The map accurately represents the proportions and dimensions of what it depicts. Yet not every map is the same. They are designed and composed differently depending on the purpose they serve. A map for a GPS, for example, will look different than an elevation map for hikers. Or a map designed to educate children will have other features than a map used for military or home decoration purposes.

The common purpose of models is to help us see and process the world from a few perspectives and give us clear reference points. Models can be used as tools for comparison, for debate, and to explain various processes such as decision-making, technical details, etc.

Mental Models

Now that we understand the overarching meaning of a model, let's put it under a microscope and zoom in on one specific kind of modeling crucial in systems thinking—mental models.

Mental models are the conscious—or unconscious —driving force behind our thinking and decision-making. We could say one's mental models equal their perspective and way of looking at the world. Some of the mental models that guide our life we learned from our parents, community, and culture. Some others we adopted by choice. And we rely on them more than we realize when we're trying to make sense of our surroundings.

The world around us is complicated and complex. We always interpret and filter data out of the stimuli our senses capture. If we hear an engine burring, even smell the odor of gas, or see the stop sign switching to red, our brain jumps to a mental model we learned in our childhood. "Cars are big. When the traffic light is red, one must stop and wait. Red means danger. The car noise means danger." So we stop at the traffic light and wait for the pedestrian light to turn green. Indeed, this entire situation unfolds in our brain in the blink of an eye. We're not consciously making this decision. Our brain already rehearsed the mental model of traffic lights so frequently that it became an automatic response. And good for us. Imagine if we had to learn over and over again what to do at a red light … We would be slowed down, and frankly, in more danger.

There is, however, a downside to ingrained mental models. Our brain is speedy to interpret known mental models. Still, it's just as quick to discard or ignore the parts of our experience that don't fit into our perception. This means we ignore or lose important learning opportunities. Besides, automatic responses led by subconscious mental models are hardly ever brought to conscious awareness, thus there is little chance to improve or challenge them. They just get reinforced in our heads for so long that they become a part of our identity. Did your aging father ever tell you, "I'm too old to change that. I've been doing it like this all my life"? My father called them quirks. Systems thinkers call them fossilized mental models.

Let's look at a real-world example. Three people, Arthur, Brianne, and Christina, are trying to decide if it is smart to invest in Facebook stocks.

Arthur is a more cautious investor. He will want to check the stock's historical performance, and Facebook stocks are relatively new. There is not much data about them. Arthur knows that Facebook appeared in the stock market in 2012, with a starting cost of $38.23. As of today, November 6, 2020, Facebook closed at $293.41. In just a bit over eight years, people who invested their money in Facebook on day one made their money back sevenfold. Not bad. But what will happen in the next few years? Brianne thinks Facebook will make significant innovations and investments and continue to grow as a tech giant. So she decides to buy stocks.

On the other hand, Christina foresees that with all the controversy surrounding the data collection of social media platforms, the days of Facebook's success are numbered. She bought some stocks back in the day, but now she's determined to sell.

Arthur, who also happens to be a data analyst on Wall Street, asks a different question, "How over- or undervalued is Facebook by investors?"

Who's approaching the question the right way? Everyone, and no one. You see, they all ask questions and make decisions based on different mental models. Of course, there is an absolute answer to how well Facebook stocks will perform, but no one can know it for sure. It depends on so many variables: employee performance, user satisfaction, ad policy changes, the market itself, the economy, the behavior of millions of investors ... So many elements weigh in on Facebook's performance. So as we said, there is an absolute answer, but it is a very complex one, and impossible to know for sure in advance. In the absence of certainty, we make predictions. How do we make the best predictions? By using models to capture different important, influential parts of the whole system.

If you asked Arthur, Brianne, and Christina to model the future of Facebook stocks, you'd get three different models. Christina would make a cluster map where she brainstorms Facebook's doom with seemingly unrelated but relevant guesses. Arthur would create a stock-and-flow model about how under- or overvaluing Facebook affects stock prices. And we would get a behavior-over-time graph by Brianne that illustrates the regular innovations and investments made by Facebook since its founding.

People create models using their own perspective, thus we get access to one piece of information when we look at one person's model. But if we put the three different approaches together, we'll get a more complex representation of reality. This is why I like to say, mental models love company. The more people work together to model a problem, the better, more complete outcome we will get.

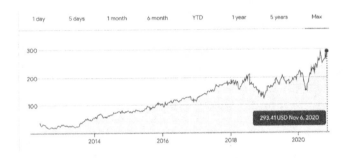

Picture 3: All-time Facebook stock market performance.

Modeling in a group setting is not always possible. And not always wanted. Say you want to draw a model where you wish to make a decision about your personal life. You want to stay as little biased as possible, but you don't want to involve other people. What to do then? Well, you can think like a modeler. What does that mean?

1. Ask good, descriptive questions. For example, should I purchase Facebook stocks?

2. State your objective. In our example's case: make more money, achieve financial independence.

3. Choose the mental model you want to use (we'll learn about a few shortly).

4. Ask clarifying questions regarding the model. What am I missing? Did I think within the right system boundaries? Is everything I included in my model relevant? Is the information I have useful?

5. Use the model to make a decision.

6. Improve the model over time based on empirical data.

So now that you have a model, you think you'll be the wolf of Wall Street? Please, don't do that. A model is just an interpretation of a larger reality. Also, models are imperfect and incomplete imitations of the real world. Just like a GPS map doesn't include every tree and rock of the land it represents, a mental model made for the stock market choice won't involve everything. A model won't give you a clear yes or no answer to almost any question, especially not to something so risky and volatile as the stock market. Those folks who claim they've "cracked the code of investing," I have one thing to say: even a broken clock is right twice a day.

Models are meant to paint a broader picture. They are useful in understanding the underlying causes of patterns in different life areas. You can assess questions about your private life, like why your partner gets angry every Christmas. You can seek answers regarding your behavior, like "Why am I switching workplaces every three years?" Or even bigger questions such as, why is the fertility rate decreasing? A model can help us investigate questions as an observer.

Where else can models be utilized? Economic growth or decline, climate change predictions, population changes, and traffic management all need models. These models are crucial pillars that aid our society and prevent chaos. As you can see, models infiltrate every area of life. Modeling is a rudimental skill to recognize, process, understand, and improve the world's problems. To become a great modeler, you have to become very good at asking one question. Namely, why?

Now we will walk through three key aspects of building a good model: define, expand, and aggregate.

Define[viii]

Imagine that someone asks you, "Do you think the policies of X political party are good ones?" What would you say? How would you define good? Chances are that what you and your inquirer understand under the term "good" is different. Let's take "a good response to climate change" as an example. A good response would involve cutting back CO_2 emissions, closing up coal factories, and switching to relying more on renewable energy sources for one party. For the other party, a good policy would be doing whatever as long as the economy won't suffer as a consequence. To avoid differing interpretations such as the two mentioned above, we must define a clear purpose for our question.

The same principle applies to models, too. One should ask a specific question, a central issue the model is meant to explore. In other words, we need to define the boundaries of the problem we want to model within the system. This defined question will be the guidepost for your model. If you reframe the issue, your model will change significantly.

After all, you'll explore different things if you pose the question, "How should we reverse global warming?" than if you ask, "What can we do today to slow down global warming?"

Thus, the defining question will lead you to draw boundaries around the elements and interconnections that are important to answer the question and to leave out the parts that are not.

This, of course, automatically makes our model somewhat reductionist, as it's impossible to sharply separate elements within a system. This is true in our personal lives, too. If you want to discover why your marriage is struggling, you can't really cut out your work life. You're working long hours and are always tired, so you get home in a foul mood and want to be alone. That will affect your relationship with your partner. The stress you feel in both life areas will affect your sleep. This will influence your work performance, your communication with clients, your quarterly assessment as an employee, and your level of self-esteem … So, where is the boundary?

The frequency and genre of our reading habits shape our understanding and outlook on the world. If you grew up in a family of academics, you were likely exposed to scientific or fine literature. This way, you develop a strong take on reading being a worthy and important activity that helps you get to know the world better, develop valuable skills, and so on.

However, if someone grew up on a farm where reading was discouraged and they had to work from dawn to dusk, their perception of reading might be that it's a waste of time. They have more limited knowledge about a good and entertaining read, why people do it, and how much they can benefit.

The less someone reads, the less they will understand the value of it. So if they had to model the question, "How much did reading help you in your career?" they'd have a very slim understanding of the concept of a helpful read. They would set their boundaries around the newspaper at best. The bookworms would make their assessment in a much wider circle and would come to different conclusions.

The point is, our understanding of the world greatly influences the boundaries we draw to investigate our choices. And our boundaries will have a significant impact on these choices. This is where our purpose comes into the picture. If we want to read quality books, regardless of how we grew up, we'll build a model to help us achieve that. If we are looking for ways to spend quality time, we will find totally different answers based on our reality. Some might choose reading and some oiling the engine of their tractor.

George E.P. Box wisely said, "All models are wrong, but some are useful." A model without a clearly defined question becomes directionless, cluttered, and unhelpful. Let's take a timely example, the US election system. If we ask how the US election works, we may end up with a tangled, complex diagram, which won't be transparent. What if we broke down this umbrella question into smaller chunks?

What is the voting policy of individual states? How does the electoral college influence voting behavior? How do individual financial contributions to a candidate's campaign impact the perception of this candidate? How does the present three-branch system often generate gridlock?

You can thus focus on the elements and interconnections that are most important to answer your question. This way, you can understand and also explain better the overall system dynamic. A defined question will lead to better solutions to fix the problem.

Exercise

Test your knowledge by brainstorming defined questions about your everyday life. How do you know you are on the right track in your relationship, work, and self-care?

First, establish a purpose you measure this question against. What additional question could you ask to narrow down the initial one (how do you know you are on the right track in …)? Write an example for each area:

- How do I know I am on the right track at work? Defining question: ………………….. (For example, how did I perform last year? Or what's the average required performance?)

- How do I know I am on the right track in my relationship?

Defining question: .... (For example, what emotions do I feel when I am together with my partner? How often do we argue? How well can we talk about our differences?)

- How do I know I am on the right track in my friendships?
Defining question: ....

- How do I know I am on the right track in my self-care routines?
Defining question: ....

The ultimate goal and purpose behind these questions is life satisfaction. We want to improve our life; it's natural. Becoming more systematic about our personal issues can be of great help.

So, where are the road bumps? What's holding you back from living the life you want? How are your life areas connected? Where could you intervene with the most leverage?

Expand[ix]

As I mentioned before, our mental models are limited because our life experience is limited; we only know what we have been exposed to. We know what it is like to be in our wealth group, race, culture, or gender. Suppose you are a middle-aged Black man from San Francisco with a six-figure income. In that case, you will have little understanding of what it feels like to be a low-income Asian woman from Texas. You think and experience different things, which shape your perspectives and, thus, mental models.

Your experience is limited to your mind. If you want to get good at modeling systems more accurately, you need to let go of the notion that you are right by default. There is depth and diversity beyond what you know. Challenge yourself by talking to people who see things differently. Read material that takes you out of your echo chamber. Seek experiences that add new value to your life.

Our way of leading life, our definition of reality, is constructed by many simplifications. Our decision-making incorporates shortcuts to make our existence more effortless. We don't really debate the wide-ranged pros and cons when we buy cereal A instead of B. For the most part, we consider flavor and maybe price. Very few people compare the ingredients and nutrition factors of cereals. Even fewer know their own nutrition needs, what vitamins they have a deficiency in, and choose food that makes up for that. A wild minority takes the effort to research who the manufacturer and distributor of the cereal support via charity. What's their political affiliation and their impact on our society or the economy? An insane amount of depth can be involved even in picking our morning crunchy. What about deciding on the bigger stuff, then? We don't even want to go there. Instead, we settle for the incomplete data we have and lead our lives based on them.

And for the most part, it works out just fine. One can lead a comfortable and happy life without seeking a more profound understanding and depth. But that's not us, right? We crave to understand more. We actively seek the intricacies of life. That's why you bought, and I wrote, this book. Our question here is, how can we become more aware of our limitations and gather relevant information to incorporate them into our decision-making process?

I think the most critical first step is that we need to leave our echo chamber. I live in one of the most liberal cities in the US. Due to the recent pandemic and social distancing, I work from home and do only the available activities outdoors. In a park nearby, there are free yoga classes every day. The park is enormous, and you can still barely find a proper six-foot radius spot to do these classes. I'm stiff as a log, but I chose to attend these classes due to no better alternative. As I got to know people there better, I became more and more convinced that yoga is a prevalent activity I somehow missed out on. People, in general, are spiritual and a bit bohemian. When I wrote the paragraphs above, it dawned on me. I'm in an echo chamber! I live in a liberal environment near a park where only yoga classes are available, and ninety percent of my social interactions besides my family happens in those classes. Of course I'm going to think that boho yoga is overtaking the world! But maybe if I go two blocks north, I'd find outdoor hip-hop or army training-like classes with people who'd never do yoga. They may not be interested in spirituality and have never heard of the concept of Meatless Mondays. Or meatless in general.

This assessment, of course, is another generalization feeding on stereotypes. There may be army training people who have a tiny Buddha statue in their homes and some yogis who eat meat with meat.

Stereotypes, by the way, are larger, more commonly accepted, mental models. They are culturally shared mental models that, whether we like it or not, do represent a little bit of truth for the majority. That's why they became stereotypes in the first place. They are our attempt to create more simplifications about the world. Stereotypes serve us well in making quick decisions but can lead us astray if we take them as the capital truth and don't investigate the problem we're facing more.

This is a challenging endeavor. The human mind is primed to seek confirmation rather than contradiction. Behavioral economists and neuroscientists proved that we operate with a bias called confirmation bias. We want to avoid uncertainty and the anxiety it causes by selectively searching for and accepting information.

Exercise

We have limited mental models—and we always will. There is not much we can do about it. We're not omniscient to possess all the knowledge that ever existed. We can, however, accept this fact and keep it in our conscious mind.

- I have limitations.
- I don't know everything.
- What if I'm wrong?
- Is the opposite also true?

These sentences are good checkpoints to stop at whenever we're about to build a mental model.

David Foster Wallace wrote a famous book named *This is Water*. In the story, he talks about two fish who swim peacefully when a third fish swims up to them and asks, "Good morning, boys, how's the water?" The two fish get confused as they have no idea what water is.

Since they were born, they lived in water. To them, water just is. They don't even think about it. The same is true of how we live our lives. Whatever your truth is, that is the water. But you can't know what water is until you experience what water isn't.

I challenge you to do just that. Leave the water. Go to places, talk to people, read literature you never did before. Try to be as little judgmental as possible. The goal here isn't to convert yourself. It is to see what other truths are there in the world and understand why those truths are "water" for other people.

Do you consider yourself a city person? Go and book a week at an Airbnb on a farm. Do you feel strongly about your political views? Go and listen to a podcast from "the other side." Do you love to read fantasies? Read a whodunit for a change! Are you rich and bored? Decide to live as the poorest in this country live for a week! The possibilities to expand your worldview are vast. Be brave enough to try something different.

Putting yourself in the way of new experiences is an essential first step. But it's not enough. The other key part of expanding your mental models is honestly doing your best to understand others' ways of thinking, perspectives, and beliefs. Also, recognize that their worldview is just as incomplete and limited as yours. Keep this in mind when you engage with others to talk about different viewpoints. Just as you get defensive when someone disagrees with you, others will likely do that, too. Avoid provocative words such as "but," "you're wrong," or "that's not true." Keep in mind that your goal isn't to convince others about your truth. You simply want to see theirs. Encourage a fluent dialogue by using words and phrases such as, "yes, and ..." "can we at least agree that ..." "I see where you're coming from ..." and so on.

If you are very adventurous, bring up your problem to people who think differently from you and ask them how they would approach this issue. Keeping their limitations in mind, become curious about how their minds work.

Systems thinkers know that everyone's mental models are limited. But they also understand that, however incomplete, each person cherishes their mental models as an important decision-making tool. When we create models for solving the larger public's problems, we need to be familiar with all these little interpretation fragments of reality to create a wholesome, satisfying solution.

Aggregate[x]

Let's stick to stereotypes a little bit longer. As very obvious physical markers, gender and racial stereotypes kick in first when we meet someone new. We can attach a lot of assumptions of someone's character if we're not mindful. This can be detrimental to the blossoming relationship, especially if the other person feels put in a box.

As we mentioned before, the positive side of generalizing is that it can make our lives easier. Making everyday decisions would be a lot harder without them. And let me tell you something you probably know but didn't think about. When we're building our models, we're making generalizations about the world. That's right. System models and mental models all rely on generalizing. The key to successful modeling is to present the model's elements at the correct level of aggregation.

What does this mean? We need to make just the right level of generalization about our elements without misrepresenting them.[xi]

Let's say you're building a model on climate change. What would be the right amount of aggregation here? We can't really put the US and Tuvalu under the same umbrella of responsibility now, can we? The first observation we make is to separate the developed and developing world populations as they have different contribution levels to environmental changes. Also, we need to take a look at the population–pollution relationship. For example, India has one of the fastest-growing economies globally yet is still considered a developing country by specific metrics. It also has the second-largest population in the world of almost 1.4 billion people. India, therefore, is definitely not your average developing nation. We can conclude that when we build our model, we also need to look at developing countries swiftly industrializing. They have a different impact on the environment than those developing countries that aren't. Our model suddenly isn't so simple, is it? Let me make it even more twisted. There are nonhuman contributors to climate change, such as cows. Cattle production is a big industry, which is responsible for significant methane gas release. This has a detrimental impact on the climate. Climate change, as a global phenomenon, is almost impossible to be modeled in one single piece. We need a ton of smaller, separate, specific models of each country with special impact. We can't merely aggregate two distinct groups of contributors, developing and developed countries. The reality is much more nuanced than that. Can you see how complex a question can get in the blink of an eye? In the case of climate change, overgeneralizations can be glaring. But they are not so obvious all the time …

How can we know whether or not we're overgeneralizing? The simple solution is self-reflection. We need to question ourselves whether or not we overgeneralized and then do some research about it.

For example, a remark such as "people who go to the gym a lot have low self-confidence" doesn't sound quite right, does it? There is this popular opinion that women and men who work a lot on their physical appearance must be compensating for something. This statement is somewhat derogatory, unkind, and frankly, it looks like a classic overgeneralization. Yet, it becomes reinforced each time someone refers to it.

(Maybe I just did that, unintentionally.) Slowly it becomes embedded into our understanding of the world and about fit, sporty people. It's not easy to free ourselves from such limiting views once they take a good hold in our brain.

What do we do about them? One can always tune down comments such as the one above using words such as "some," "a few," and "certain." "Certain people who go to the gym a lot have low self-confidence" sounds one beat better. But this kind of phrasing proposes another issue, naming who we are referring to. Once we become more specific in our generalization, we may disaggregate so much that it would make the entire situation odd.

It is better to avoid overgeneralizing statements altogether. What we can do is be mindful of our thought patterns, and before we decide to share our observations with the world, we can test it internally: "Is it true?" We don't have to think for a long time to realize that not every fit person is insecure. We can come up with a lot of people who aren't. Similarly, does it mean that less sporty people are more confident? Not necessarily, right? We'd soon end up saying "all people are insecure," which is just a ridiculous statement, but that's the logic of overgeneralization.

Exercise

Let me be the Devil's advocate here. Try living your life for a day in the land of wild overgeneralizations. Whenever you come across a stranger, try to come up with the most ludicrous overgeneralization you can. Don't tell them, of course, but entertain the ideas in your head and feel how untrue they actually are.

Here is a targeted practice. Soon I will tell you a few words. These words will describe people either by their profession, race, political affiliation, physical appearance, and age. Your task will be to think about any stereotype—positive or negative—these people attract:

- bankers

- rich people

- bosses

- surfers

- writers

- surgeons

- homeless people

- Black man in a hoodie

- middle-aged short blonde-haired white woman

- old Asian woman

- fat child

- Republican/Democrat

- tall people
- short people

Did you attach the trending stereotypes to each of them? Now, let's make it more interesting. What are your personal stereotypes of each of them? These may or may not match popular stereotypes.

Okay, now comes the exercise for some of the bravest. Share some of the overgeneralizations you collected with a trusted friend or acquaintance. Don't tell them that you're completing an exercise from this book. Just pretend that the stereotype is your opinion, and you want to engage in a conversation about it. How does your friend react? Do they engage and agree, or do they point out that you're using a stereotype? You can inform them that you just wanted to try out an exercise from a book at the end of the conversation. This confession may trigger a more in-depth discussion.

The next day flip the coin. Do your best to avoid stereotypes at all costs. Also, be on the lookout. If someone else says one, point it out. Ask these people who exactly they know personally who fits their comment and whether they genuinely believe in the stereotype. Correcting someone's overgeneralization can derail a conversation and shift the mood. How does the person put on the spot react? Annoyed, angry, defensive, or grateful? Being in the role of the stereotype police is actually a thankless job.

A good modeler keeps their assumptions in check and knows that both under- and over-generalizing are detrimental. One needs to balance them well for each problem they investigate.

To close the stereotype topic, I have a final practice for you. I will share again the list of people to who you attached some kind of stereotype:

- bankers
- rich people
- bosses
- surfers
- writers
- surgeons
- homeless people
- Black man in a hoodie
- middle-aged short blonde-haired white woman
- old Asian woman
- fat child
- Republican/Democrat
- tall people
- short people

Now your task is to challenge these stereotypes. Ask yourself the following questions:

- What basis does this stereotype have? Why does it exist?
- Who benefits from maintaining the stereotype?

For example, the stereotype of a douchebag boss benefits dissatisfied employees. This stereotype gives them peace of mind, an excuse even as to why their life is unhappy. "It's my boss, that typical bully. It's their fault." And don't get me wrong, their boss may be a douchebag. But stereotyping derails them from the broader question: "Why do I tolerate this? If it is so bad, why don't I make a change?" The mean boss archetype is also an escape from taking responsibility for their life. A more honest and self-aware response would be either taking the risk of changing their job or admitting that they are afraid to make the change. In both cases, they reclaim responsibility.

Answer the original question, who benefits from maintaining the stereotype, based on my example.

- Do I personally know someone who fits into the stereotype in question?
- Rewrite each stereotype, taking away its generalizing power. (For example, bankers are usually considered greedy, money-driven, and such. You can use the softening word some. "Some bankers are greedy and money-driven … but so are so many other people in other professions." Even teachers can be greedy. Have you seen *Breaking Bad*?)

Systems Mapping

Now that we got acquainted with how mental models work and how we create them, let's zoom on to our next topic, namely, system maps.

Systems mapping is a tool to help you visualize the system you are analyzing. It requires an understanding of the elements, their functions, and their interrelationships within the system. Knowing all this, you can start to chart and see how you can intervene and influence the system in the best way possible.

Let's recapitulate what we talked about in the previous chapter regarding systems thinking so far:

•   Interconnectedness shows how things are connected.

- Synthesis shows how two things can combine to create something new.

- Emergence shows how self-organizing elements come together.

- Feedback loops show the dynamic and direction of the behavior of the elements.

- Causality shows how one thing can influence another.

- Systems mapping gives an overall visual representation of how elements interact with one another within the system.

Why Learn about Systems Mapping?

Because it is a crucial component in making large problems more transparent. It is essential to pin down and accurately assess relevant leverage points where our interference would bring the greatest benefits.

Today the world is burdened by many challenges like a pandemic, climate change, and other natural disasters. We often pinpoint the exact causes of these challenges and like to alleviate the symptoms they produce. This is linear thinking, right? It is reductionistic and makes us only focus on isolated parts of the problem. However, these massive challenges are caused, maintained, and suffered by all of society. Isolating a single part of the challenge isn't going to solve the bigger problem we are facing.

The following chapters are going to reveal the most commonly used systems mapping tools. We will learn about cluster maps, interconnected circle maps, causal loop diagrams, behavior-over-time graphs, and stock-and-flow diagrams.

Chapter 3: Cluster Maps

What Are Cluster Maps?[xii]

A cluster map has quite a self-descriptive name. You dump every idea your brain can come up with concerning the complex problem you're dealing with. This is not a place to be selective about your brainstorming. Just allow a free flow for your thoughts. Keep in mind, the ideas have to be related to the system (complex problem).

Visually speaking, in a cluster map, we write the central problem, question, or topic in the middle of a paper. Then, around this central issue, we write our other related ideas. These ideas will serve as the elements of the particular system we're exploring. We can write them wherever on the paper as they are interconnected. Once we've collected every element we could think of, it's time to discover the interconnections. This will be the part that reveals the not-so-obvious aspects of the system. It will significantly help us to expand our view and understand the actual complexity of what we're dealing with.

The more data we collect from mapping out the interconnections, the more we will see which parts of the system need our intervention the most. This being said, compulsive and obsessive intervening will not lead to better outcomes. Like aggregation, we need to find that fine line where the intervention will benefit the overall system dynamics. Some people want to interfere at all costs before they've properly understood and embraced the system's natural chaos.

Any type of relationship is messy. And so should your cluster map be. The more chaotic your map looks after you connect everything with everything, the better. On the one hand, you will have a chain of insights, one leading to the next. On the other hand, some elements will start to stand out based on how many connections they have. The more "roads" lead to a proverbial Rome, the more central role that element plays in the system.

There are no right or wrong answers in a cluster map. The entire experience is just one big brainstorming session. There are system maps that require engineer-like precision. This is not one of them. The goal of a cluster map is to understand better the area you're exploring.

How to Create a Cluster Map

1. Take a piece of paper and pens of at least five different colors. The more colors you have, the more transparent and organized your cluster map will be. You can associate a particular collective idea around a color—for example, red stands for close relationships, blue stands for opposing views, and so on.

2. Now, the most critical part. What are you exploring? What's your central theme, question, topic, or problem? Write it in the middle of the paper. This could be anything from childhood trauma to your hometown's development under the current mayor.

3. If multiple people participate in the cluster mapping practice (which is fantastic), make sure that everybody has access to a pen. Everyone should add their ideas. If your group is more than four people, I would suggest a large A3- or A2-

sized paper to brainstorm over so people can have easy access to it.

4. Remind everyone—or yourself—to write down any and every related idea to the topic. What's the worst that can happen? The idea (element) will get no interconnections in the later phase of the practice. Sooner or later, only the truly relevant elements will stay in the game, so to say. Release your urge to find associations when you explore the elements of the system.

5. When no new idea is generated, it's time to map out the interconnections. Use the colors here to represent different flows.

6. Continue connecting elements until you feel there's nothing left to connect. It's important to speak out loud about why you make a connection. For example, A is connected to B

because ... and describe their interaction. Especially if you are working in a group, everyone needs to understand your reasoning behind a connection. Encourage everybody to explain their viewpoint for making a connection between two elements.

7. Once you have some interconnections, brainstorm new insights about the problem thanks to the cluster map. Have a group discussion about your findings; are they relevant, do they change anything, and how can they be used to improve the system?

Example of a Cluster Map

I will not beat around the bush. Here is the cluster map of Ed, a senior manager at a marketing firm. Ed complains about hypertension and feels stress constantly. When he learned about cluster maps, he decided to take a closer look at his life.

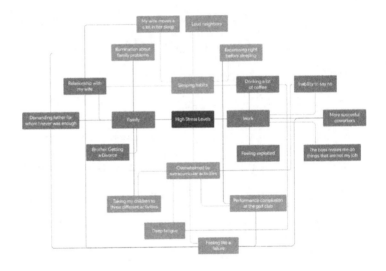

Picture 4: Ed's cluster map.

Let's break it down. As you can see, in the middle, we can find Ed's high stress level, the core problem we're investigating. Ed generally considers himself a well-rounded person: calm, successful, blessed with a great wife and kids. He has money, a good job, hobbies—he shouldn't be so stressed on the surface. But as we know, the devil lies under the surface. So, Ed decided to explore where his undesired emotions might come from. (Follow the story I tell below on the cluster map above. I describe the thought process that created this cluster map and its interconnections.)

After writing down his core issue, high stress levels, in the middle of the paper, he divided his life into four major areas: family, work, extracurricular activities, and sleeping habits.

He starts his brainstorming with his family. He has a great relationship with his wife, but sometimes they argue. He has a not-so-great relationship with his dad, an overachiever who was never satisfied with Ed. His brother is going through a divorce, which does affect Ed's mood, of course. He needs to give a lot of his attention and free time to his brother.

Ed quickly realizes that he actually doesn't feel that great about his free time (the little he has). So, he adds "overwhelmed by" to his extracurricular habits. Ed feels a lot of pressure to perform at the golf club. He doesn't even understand why peak performance is so essential there … Maybe because he doesn't want to feel like a failure. Yes, he definitely doesn't want that. Man, he just feels so tired every time he has to do something after finally getting home from work. Profound fatigue is definitely something he associates with free time. Oh, and the kids' extracurricular activities—piano classes, football games, a French teacher, ballet … There is no end to the chores.

Talking about chores, Ed turns his attention to the work bubble. No doubt, he has lots of duties at work, too, as his boss consistently dumps tasks on him that are not even his job! Ed jots this down in a bubble. And him not being able to say no (another bubble) always has him ending up with the shorter end of the stick. His boss, his coworkers … they all ask him lots of favors. He actually feels exploited (another bubble) … No wonder his coworkers are more successful than him (another bubble). They can focus on their advancement while Ed does the slave labor for everybody! All he does is drink coffee and work. Repeat.

It dawns on Ed while he fumes about his work misery … No wonder he can't sleep at night. He needs to chug a double black sometimes at 6 pm. Drinking coffee is a bubble that is clearly connected to his job, but he can just as well attach it to his sleeping habits. Just as feeling fatigued is related to drinking coffee and his inability to say no affects his free time, where he also is a yes man. So many things draining his energy …

No wonder he is so sleep deprived. He could connect sleep deprivation to the deep fatigue bubble he put in his extracurricular activity cluster. And this ties directly with his coffee consumption, as well. It makes sense. It's a vicious cycle. He can't sleep properly, so he doesn't have enough energy to meet the day's challenges. He feels fatigued, so he drinks coffee, which messes with his sleep. This creates a reinforcing feedback loop in his life. (We will learn about it more in the next chapters.)

He is a light sleeper anyway. His wife likes to run marathons in her sleep, kicking and pushing

Ed now and then. Also, his neighbor is partially deaf, so he listens to his TV very loudly. He puts both of these observations into a bubble.

To be fair, he argues plenty with his wife about his inability to sleep. He is aware that his wife is not at fault for being an active sleeper, but sometimes he doesn't care. So he bickers. He can see that his sleeping misery creates some tension in his family life (see the "my wife moves a lot in her sleep" and "relationship with my wife" connection).

Another thing keeping him awake at night is the endless rumination about what he could have done better to please his father. He can definitely find a connection between his low quality of sleep and his father's weekly check-ins. His father asks each time if he finally got promoted to vice president at the company he's worked at for over 10 years. This question makes him feel like a failure (he connects the dots between the "demanding father for whom I never was enough" and "feeling like a failure" bubbles). The choking belief of not being enough spills over into other areas of his life. It affects his relationship with his wife as he barely ever stands up to her. Ed also feels discouraged by his colleagues' success, which further enforces his insidious feeling of inadequacy.

When the cluster map is done, Ed takes a deep breath … He just made a significant discovery. Many elements mentioned on the cluster map are the symptoms of his high stress level. But two outstanding elements affect most of his life areas. These are the bottleneck points that are the actual source of his suffering. Can you guess which two?

One is his inability to say no. He can't say no, so he engages in more extracurricular activities than are pleasant. He can't say no, so his boss overworks him. He can't say no, so his colleagues build their careers on his back ... No wonder he feels like he is not in control of his life. Feeling exposed to the circumstances can generate a lot of stress.

The other key issue is his general feeling of not being enough. It started with his relationship with his father. This insidious feeling now runs many areas of his life: his romantic relationship, work, and even his hobbies. He must win to avoid feeling like a chump. Trying to perform at his best, driven by bottomless perfectionism, is also a crucial stressor.

Ed was suspicious about a few things in his life, the loud neighbors, the overtime at work, his father's nagging. But before the cluster map, he couldn't understand his emotional make-up—and his real stressors—this clearly. Now he knows where he needs to focus on his life to diminish his stress.

Exercise

Now it's your turn! You saw how a cluster map works—time to create your own. Choose a problem very central to your personal life or something global such as the brain drain of poor countries.

Chapter 4: Interconnected Circle Maps

What do you think? Is an aggressive approach to foreign policy more likely to deter war or to start one? Are pesticides that keep crops healthier and harvests more abundant worth the risk of an ecological disaster? Have you ever thought about exploring a negative movie character's actions from multiple perspectives?

Some questions appear quite simple on the surface, but are they really? And the answers to these questions are rarely black or white. These issues occur on a spectrum of gray. Those who practice systems thinking know that identifying the central causal relationship within a problem is difficult. Finding the bottleneck point of optimal intervention is even more challenging.

An interconnected circle map can be a valuable visual tool to explore the problem areas of a system. It can give the analyzer the potential to dive deeply into the causal relationships of the dynamic system. Interconnected circle maps expose the parts within the system and show the whole picture at the same time. One can synthesize the most significant flows in the system. Also, one can discover divergent elements and develop an impactful intervention plan to fix the identified issue.

An interconnected circle map is a great collaboration tool. Like cluster maps, they are best done by a group of people where many observations and opinions can be represented.

Like the progress from level one to level two in a computer game, that's how cluster maps relate to interconnected circle maps. How can you transform your cluster maps into its advanced interconnected circle version?

Document every significant element in your system on a piece of paper and create a cluster map just like you learned in the previous chapter. Then cut these out from the paper so you get little tags that you can move around. Draw a sizable circle on a separate piece of paper. Then position your little tags around this circle, outside of the periphery. This way, you can edit, reposition, and synthesize the parts. Based on a discussion with your group, connect the elements that have a relationship with one another, mapping out the system's dynamic patterns. As you progress with the interconnections, the more significant and less significant elements will reveal themselves. Using this information, you can start debating which part of the system is responsible for the feedback mechanism you want to alter.

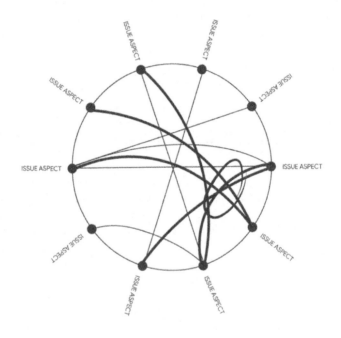

Picture 5: Interconnected circle map.[xiii]

This simple thinking tool is great for illustrating complexity. Here we can share our ideas about the changing behaviors within a system, pick the most significant elements in light of those changes, and trace cause–effect relationships with arrows. This way, we preserve the complex nature of the system (everything outside of the circle periphery) and can see and manage a variety of interactions at once (everything within the circle).

If cluster maps are level one in a computer game, interconnected circle are level two. And level three would be causal loop diagrams. I will talk about causal loops in detail in the next chapter, but here I would only like to mention that interconnected circles can be a helpful jumping board to make causal loop diagrams. The challenge in designing a causal loop diagram often is finding its starting point. An interconnected circle map will already contain the principal elements and significant connections, so detecting change-driving feedback loops afterward won't be that difficult.

Step-by-Step Breakdown[xiv]

Let's have a look at the steps necessary to create your very own interconnected circle map.

1. Choose your topic, research area, or story you want to explore deeper. The more change happens over time, the better. You can literally use anything as your source to choose your topic—the latest news, a book, or your own diary logs. Start with something simple. Recently, I got into the show *Breaking Bad*. It is an elaborate exhibition of character development packaged in the grim topic of drug-related activities. I warn you now that my brainstorming may reveal some spoilers if you are early in the series. Of course, I will not tell you the ending, and I will try to be as

general as possible. In my story, I focus on the character development of the main character, Walter White, a high school chemistry teacher turned into a meth manufacturer.

2. Simplify the story of your choice. Interconnected circles are meant to help us understand complex topics. However, some aspects of our story, like the choice of jargon and content, may be problematic for some people to follow. Also, most stories and articles are lengthy and tend to elaborate on things that are not important from our perspective. Talk about and define the central problem—the main focus of your research. In my study, I chose to focus on the character development of Walter White. Every other unrelated (yet super exciting) event in this story will be ignored.

3. If you work on this exercise as a group, divide it into three to four smaller groups depending on your number. This step is not necessary, but research shows that collaborative brainstorming improves the thinking process, the flow, and overall conclusions. Everyone should be familiar with the article, topic, or problem you're about to discuss. Make sure that each person has access to the initial data. In my case, I sat down with a group of friends who all watched *Breaking Bad*.

4. Distribute a sheet of paper with a big circle on it (the future interconnected circle) to every participant.

5. Start thinking about a couple initial elements that are crucially important in the case you are studying. Include no more than five to ten elements. Write them outside of the periphery of

the circle. The elements of the story should satisfy the following criteria:

- They show important changes in the problem.
- They should be nouns.
- They add or subtract from the problem.

Additionally:

- You can try to find elements that cause the change (growth or decline) of other elements.
- To illustrate the cause–effect relationship, use arrows from one element to the other.
- Only direct causal connections should be represented on the map.
- Search for feedback mechanisms.

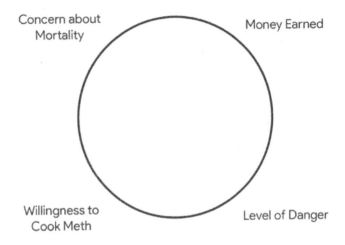

Picture 6: The Interconnected Circle of Walter White

6. Everyone in the group should draw their own interconnected circle, even though they will discuss the group's elements. Encourage open communication between members about their opinion regarding the topic. If there is disagreement among people within the group, everyone should be allowed to write their own

element ideas. In any case, as the model gets more and more refined, it's possible to change, erase, or add elements to the circle.

What matters most here is the cognitive involvement and information processing, not just the outcome. Ensure that people are focusing on their central thesis or question. They should be specific in their word choices and intentional about finding the "change" that creates the problems.

Be precise with labeling your elements. In my case, first, I put "Cooking Meth" as a critical element of Walter's choices, but then I realized it was too vague. So I rewrote it to "Willingness to Cook Meth." The amount of meth he cooks is not so important from our main perspective: the personality change he goes through thorough the show. His choice of whether or not to cook is more meaningful.

Similarly, "Mortality" is the main element—he discovers that he has terminal lung cancer. But this label would say nothing about the motivations of Walter. He jumps into the meth business because his mortality concerns him. He is worried that he'll leave his family—his ill son and pregnant wife—with nothing. Therefore, "Concerns about Mortality" is a better way to describe the element.

As you can see, the two elements I talked about so far are intangible; concern and willingness are measurable concepts. The "Money Earned" element, on the other hand, is a tangible one. Remember this, elements can be both tangible and intangible.

7. While everybody has their own interconnected circle map, you should have the main one, maybe one drawn on a whiteboard in the middle of the room. Everyone should work together on this one. After some time has passed and everybody has discussed their element theories among their peers, you should ask each group to suggest elements. Tell people that it is completely fine if they change their initial ideas based on new information they hear from other groups. Agree upon and select up to five to ten elements to write on the main whiteboard.

8. Start mapping out the causal connection between the elements. First, allow people—and groups—to do their own work. Then, just like before, ask groups one by one to suggest connections between two elements on the whiteboard.

While someone describes the connection, make sure that they mention if one element triggers an increase or decrease in another element. For example, the higher Walter's concern about his mortality, the higher the likelihood of being willing to cook meth. But as the level of danger around his new business increases (the DEA starts sniffing around him and competitor drug dealers become protective of their territory), his willingness to cook decreases. And, of course, the money he earns moves in the same direction as his willingness—and action—to cook.

To illustrate these observations, draw arrows between the appropriate elements. (See Picture 7.)

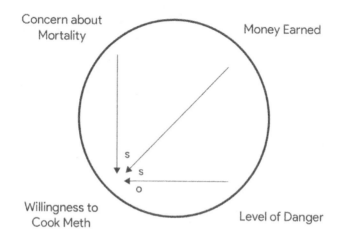

Picture 7: Walter's Decision Making.

9. Highlight that one element can have multiple connections and encourage people to try to find some. Some elements may not be connected to anything. Everyone has to be able to explain why they connected two elements and what dynamic they discover between them. For example, if someone connected "Level of Danger" with "Concerns about Mortality," it is an exciting

proposition. In the beginning, Walter White was only afraid that cancer would finish him. But as the show evolves, he is gradually more terrified of other things that could kill him (some cartel members, for instance). There is definitely a relationship between those two elements.

10. Look for closed loops. In other words, can you connect element to element and end up at the starting point? There can be multiple closed loops within an interconnected circle map. Usually, the more elements the loop has, the higher the chances for numerous closed loops.

Each closed loop should be marked with different-colored pencils. One has to be able to tell the story of each of the loops. For example, Walter White was concerned about his mortality, so he was willing to cook meth to make more money. Once he has more money, his concerns about mortality decrease. (This is Loop 1.) However, the more willing he is to cook (and sell) meth, the more his competitors want to kill him, so his mortality concerns increases. (This is Loop 2.)

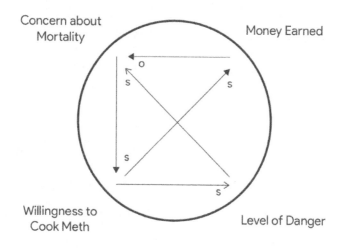

Picture 8: Walter's Closed Loops.

11. Assign a feedback structure to your loops. In my example, loop 1 is a balancing loop; Walter's

mortality concerns are high, so he starts to cook meth. As a consequence, he earns more money, so his concerns about his mortality decrease. Loop 2 is a reinforcing loop: Walter is willing to cook meth, which earns the cartel's wrath. He gets death threats from the shady elements, so his fear of death increases. He grows more and more impatient to scale up the business to make money to fulfill his goal, to leave his family with enough when he dies.

12. Once everyone finished working on their personal interconnected circle, it's time to share what each person—or group—discovered. Are there any common conclusions? Ask people to describe their method of finding their feedback loops and the changes these cause in the story.

Your Turn!

Based on the 12 steps I presented above, create your own interconnected circle. Choose whatever topic you wish. Do the practice alone or with friends. But most importantly, practice doing it. In the following paragraphs, I will provide further guidelines to help you make the best deductions based on your map.

- Were any elements more "popular" than others? Did some element have an unprecedented number of connections? The elements that attract or dispense many arrows tend to present the area where intervention is needed. These elements are responsible for a lot of change. In my example, a lot of the key issues revolve around Walter's concerns about his mortality. That is the fundamental drive that set his other choices into motion.

- Are there any elements that don't have any connections? Are they truly important, then? Is there something missing that could connect this element to the flow of the circle? In my example, there were no unconnected elements. But I could have added some. For example, some old friends and business partners offered to cover Walter's hospitalization costs. But he (SPOILER ALERT) declined their offer. In the first episode, it is unclear why, but as the show goes on, we learn their relationship's background story. So, "Readiness to Accept His Friends' Help" could be a standalone element, yet an important one. As we progress in the show and understand the friendship dynamic, we could add an element to the story that would connect Walter's "Readiness to Accept His Friends' Help" to the circle.

Whenever we don't have arrows pointing towards or coming from an element, we need to reassess our story. Some additional elements may be missing.

- What's the significance of closed loops within a circle? This means that we found a feedback loop. When one element of the feedback loop is altered, its effect will ripple through every element involved in that loop. For example, in Loop 1, the starting point is Walter's increased concern about his mortality. He is worried about his family's financial security. He hasn't got time to leave them with a secure financial background. This fear prompts him, the model citizen, to start doing illegal things (cook meth) to make money quickly. As this business starts rolling and makes more and more money, his deep fear of leaving his family with nothing eases. But, as we saw in Loop 2, his peace of mind is short-lived as his actions bring other elements to the table—the cartel, in this case. Walter is chased and threatened by them, so he must work faster, gain a bigger "market share," for a bigger profit. With each batch he cooks, he gets involved deeper with the drug world. This is all the doing of the reinforcing Loop 2.

- Track the loops and see what happens to each element involved. If we check Loop 1, we can see that Walter's health concerns and willingness to cook move in the same direction. Ceteris paribus, if he was unbothered by anyone, he could have cooked the right amount of meth to earn the amount of money he initially planned to make. The more money he had hidden in his closet, the more his concerns about his mortality would have decreased. Again, hypothetically, when he reached the sum he wanted to save up, he would have arrived at a place of peace regarding his mortality. Theoretically.

- But life didn't work out in his favor. If he has a smooth ride, there wouldn't have been a five-season TV show about it. As you could see, some elements were part of both Loop 1 and Loop 2. That can often happen. It means that the loops interact with each other, which makes our story more complex and exciting. As we could see, depending on how Walt gets stimulated, his willingness to cook meth can increase or decrease over time.

The final goal of creating interconnected circles is not to find a final solution to problems. Instead, it is a thinking tool to better understand the system's story, brainstorm new ideas, and clarify the complex problem's main catalysts. We can highlight what is changing in the system and trace the intricate connection pathways to understand those changes.

Interconnected circles are a great jumping board for designing more complex system maps such as causal loop diagrams or stock-and-flow diagrams.

Chapter 5: Causal Loop Diagrams

System dynamics is the sum of activities happening within a system. These behaviors create feedback loops that balance, maintain, or change the system. We can better understand these processes if we broaden our perspective on causality, namely, how one action leads to another in a dynamic, ever-changing system.

We previously learned that there are two kinds of feedback loops, reinforcing and balancing. In a reinforcing loop, the elements repeatedly perform the same action, resulting in exponential growth or decay. In a balancing loop, the elements of the system balance each other out.

Both loops can bring along positive or negative consequences. A balancing loop can cause stability or stagnation. A reinforcing loop can be responsible for desired or undesired growth (or decay).

Causal loop diagrams (CLDs) are meant to give us a conceptual framework and language to understand the reason behind balancing and reinforcing dynamics. We can see each diagram as a sentence with the elements as subjects, the interconnections as the verb, and all the other syntax parts to explain the causal relationship between them. Adding multiple sentences (diagrams) together, we get a coherent text; we end up with the story of the system's problem. A causal loop diagram chain basically shows how things interact.

This chapter presents how to create causal loop diagrams, which represent feedback in a system. These diagrams are great for:

- "Quickly capturing your hypotheses about the causes of dynamics.
- Eliciting and capturing the mental models of individuals and teams.
- Communicating the important feedback processes you believe are responsible for a problem.

The conventions for drawing CLDs are simple but should be followed faithfully. Think of CLDs as musical scores: At first, you may find it difficult to construct and interpret these diagrams, but with practice, you will soon be sight-reading. In this article, I present some important guidelines that can help you make sure your CLDs are accurate and effective in capturing and communicating the feedback structure of complex systems."[xv]

The Process of Building Causal Loop Diagrams

1. **Select a guiding question/problem/theme.** Designing a causal loop diagram without a guiding question is like trying to use a GPS without having a destination—it's pointless. The goal of a CLD is to help you see deeper insights into the complex problem you're facing. For example, "understanding how the cheaply manufactured Chinese products affect the market of the more expensive Western manufacturers" is a better problem definition than "understanding market dynamics."

2. **Timeline.** To create well-rounded CLDs, it is helpful to impose a time interval for our guiding

question. This interval should be generous enough to allow our diagram to show how the dynamics play out. But it should also stay relevant. We don't care about the Western–Eastern market dynamics of the 16th century, but it can help to take a look at the distribution change of manufacturing patterns from the Second World War onwards.

3. **Observe the behavior over time.** There are specific charts to help us do this analysis, and we'll learn about them in detail in the next chapter. Here, I'd only like to mention that understanding how the elements behave over time is crucial to make more accurate predictions for the future. Making such predictions is undoubtedly risky and often inaccurate. But we still do them because we can test our assumptions and identify inconsistencies that we would overlook otherwise. Behavior-over-time graphs

also help us notice important elements that we may need to include in our causal loop diagram.

4. **The question of boundaries.** Previously I said that a behavior-over-time graph can help us see new and needed elements. But when do we know we have enough? It's quite easy to be derailed from the initial guiding question and add elements that overwhelm your study (and you). Keep in mind that you're not trying to capture the entire system, just the guiding question. A good rule-of-thumb question to ask yourself whenever you're in doubt is, "If I doubled or halved the element, would it impact the guiding question's central problem significantly?" If the answer is no, you can probably skip that element.

5. **What's the right level of aggregation?** Usually, the guiding problem itself imposes the need for complexity. Some issues, even if broken

down, need more details than others. Generally speaking, the variables should not talk about specific events (feeling anger). They should illustrate behavior patterns (feeling angry every time you speak to your father).

6. **Look for delays.** Some interconnections may operate with significant delays compared to the general dynamic in the system. Finding these delays is needed because they can be the reason behind systemic imbalances. For example, imagine if a manager expects returns on investment immediately after releasing product Z. But he fails to observe a delivery delay, a processing delay, and a payment delay involved in product Z's selling process.[xvi]

Now that we've taken a brief look at what important aspects we should look out for when building a causal loop diagram, let's look at how to actually do it. Following Daniel Kim's guidance, let's go through some tips for successful causal loop building:

1. Use nouns to describe your elements. Avoid using verbs. For example, use "Population" instead of "Growing Population." When there is a decreasing tendency, you would have to refer to your element as a decreasing "Growing Population." It's weird. The links you will use to show the connection between two elements will specify if we are talking about a decreasing or growing population.

2. Use measurable things as elements. Shifts in the level of "sadness" can be measured, but "mental state" can't.

3. When it applies, use the positive aspect of a state. And an increase or decrease in growth is

easier to comprehend than an increase or decrease in decay.

4. Whenever you present an action in your diagram, make sure to explore both the intended and unintended consequences related to it. For example, an increase in "family time" may increase the "quality of your relationship with your children." But it may decrease your "work performance" and your "time dedicated to working."

5. Make a clear distinction between perceived and actual states. The perceived quality of a relationship may be different than the actual.

6. If there is more than one relevant consequence of an element, merge them into one phrase. For example, due to decreased time dedicated to working, you start rushing, deliver less value,

and have many grammar errors in your assignments. You can commonly refer to these as "work assignment quality."

7. When the connection between two elements is hard to explain or is not obvious, there is often a need for an intermediary element. For example, we might not understand how "family time" is linked to "work assignment quality." But if we introduce the element of "time dedicated to working," that explains the relationship better.[xvii]

Generally speaking, when it comes to causal loop diagrams, less is more. The goal of this diagram is to illustrate the problem simply. You can add more items to the diagram as you progress in the story, but start by presenting the problem with the most necessary parts. A standard diagram may be sufficient to jumpstart the brainstorming session about alternative paths for looking at the issue. If the issue is more complex, it's better to use multiple simple loop diagrams to illustrate the interrelationships between the system elements. Complex doesn't need to mean complicated.

The best causal loops show the connections and relations between the system elements and highlight the aspects we are not conscious of. Loops are the simplified representations of the present. Let's take a look at a visual example.

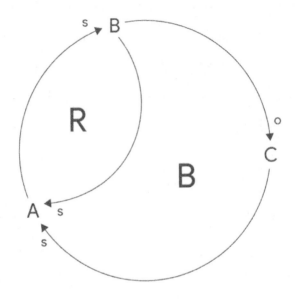

Picture 9: Causal loop diagram.

Causal loop diagrams are mental models that serve us to present cause-and-effect relationships within the system. There are, however, some less understood, unspoken processes behind these connections.

How does a change in a company's reward system affect a worker's performance? How does a shift in the number of funds available for research and development affect the discovery of new technologies? How does your lack of self-care affect your health later in life?

Taking a more in-depth look at the links presented in such cases can help us become more aware of the structures that create the observed behavior. We then can find more fitting solutions to improve.

Going Deeper is a process in systems thinking where we add thought bubbles to causal loop diagrams and explore the not-so-obvious parts of a system.

When we go deeper, we first create a causal loop diagram to illustrate the system's problem. We look for connections that are made by human choice (versus rigid physical structures). For example, suppose we have a connection where a change in weather conditions affected the crops. In that case, we're facing a problem created by nature. But let's analyze a connection between a change in governmental funding and investment in cancer research. That's a classic case of human choice.

After you pin down a few connections that are considered human choice, you can start digging deeper, asking questions like, "Why was this choice made?" To physically illustrate the implicit process of thinking, we attach a thought bubble to the connection.

"What do we put in the thought bubble?" you may ask. Ideally, you (or your group) try to position yourself in the situation. You can use think tank methods, role plays, try to empathize—whatever works. The thought bubble should include the rational thinking path of each of the elements acting in the loop.

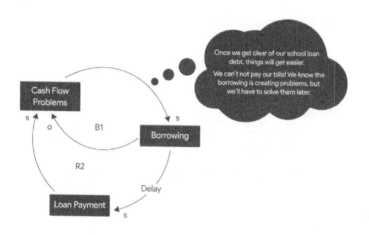

Picture 10: The causal loop of borrowing money.

Let's say you and your family face financial difficulties, so you decide to borrow money from your bank. (See Loop B1 in Picture 10.) The loan covered your immediate needs, but you didn't develop any new revenue streams. To pay back your loan with interest, you need to take out another loan. (See Loop R2.)

How to Design the Causal Loop

In our borrowing example, as the cash flow problem increases, so does the borrowing problem. Then a balancing event happens: As your money increases thanks to the bank, the cash flow problem decreases—for a bit. This is what you see in Loop B1. As time goes on, you need to start repaying your loan, and cash flow becomes a burning issue again, as you can see in Loop R2. This situation is a classic systems archetype story called "Fixes That Backfire."

The thought bubble is added to the connection that represents a human choice. In our case, human choice is fixing the cash flow problem by borrowing. This is why we added the bubble to the line connecting "Cash Flow Problems" and "Borrowing." Assuming that we act rationally when we decide to borrow, we should ask ourselves why we got the loan. Maybe we want to stay afloat, for now, hoping for a better future—a promotion, inheritance, paying off the debt with the highest interest rate, etc. We could also be acting out of desperation—we are hungry now! We know another loan won't solve our long-term problems, but the short-term issues are more urgent. We could be making an emotional decision, such as not wanting our children to worry by noticing a lifestyle change.

Focus on multiple perspectives when you fill out the thought bubble. Perhaps you're looking forward to that long-promised promotion. Your wife may be looking forward to going back to work after an illness. Capturing multiple positions helps you have a more accurate grasp of the situation.

The point of Going Deeper is to detect and pinpoint those not-so-obvious reasons for behaviors that are not on the surface. Jumping to premature judgments like "this is so irresponsible" or "whoever does that has no financial education" will provide hasty conclusions and unhelpful solutions. It's always better to learn about every individual problem a few layers deeper than the obvious.

How to Determine Whether a Loop Is Reinforcing or Balancing

John Sterman, a contemporary systems thinker, distinguishes two ways to assess whether a loop is reinforcing or balancing. He refers to these as "the fast way and the right way."

The fast way is to count how many negative (−) or opposite (o) links are in the loop. If it's an even number, the loop is reinforcing. In the case of an odd number, the loop is balancing. But as we mentioned before, polarity-labeling errors can happen within the loop, in which case the number of negative links could be misleading.

According to Sterman, the right way is to make a small change in one of the loop elements and see what effect that change creates.

Trace the effect of a small change in one of the variables around the loop. Choose any element in the loop and increase or decrease it in your mind. How did this change affect the loop's dynamic? If the change introduced by you just builds upon the original dynamic, we're talking about a reinforcing loop. If your change opposes the original dynamic, we have a balancing loop. This method proves to be correct regardless of the labels attached to the links.

Take another look at Picture 10. Let's trace the two methods I presented above. With the negative or opposite link counting method, we would answer that the two-element loop, "Cash Flow Problems" and "Borrowing," is balancing because it has an odd number of negative or opposite links. The three-element loop has no negative or opposite links (zero is considered an even number here); therefore, it is a reinforcing loop. The labeling on the diagram looks correct, but what if I made a mistake?

Let's check the loops' nature based on Sterman's right way. Let's choose an element. I choose "Borrowing." Say we increase the amount of money we borrow by 30 percent. The increased amount of loan would temporarily decrease our cash flow problem. The decreased cash flow problem would lead to a diminished need for new loans (remember, we're talking about the short term here). Our first loop is balancing because the new change we introduced is contrary to the original dynamic where our need to borrow money was increasing.

In the three-variable loop, I choose to increase the loan payment. For example, the borrower signed for a variable interest rate. Due to economic changes, his loan repayment amount grew significantly. The heavier repayment burden would cause even more cash flow problems (just like in the original dynamic). The more pressing cash flow shortage would demand even more loans, which would result in even more high-interest loan repayments. Changing the element in this loop just reinforced the original dynamic; thus, we're talking about a reinforcing loop.

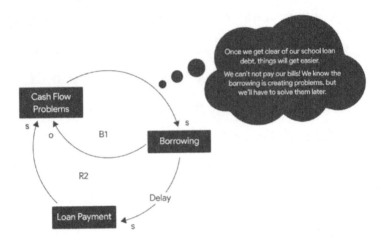

Picture 11: The causal loop of borrowing money.

Avoid Link Ambiguity

Links can be positive (+) or negative (−), depending on the dynamic between the two elements it connects. For example, if we draw a diagram where we add two elements, a business's profit and its products' price, the link between them could be either positive or negative depending on the products' demand. For instance, the company decides to raise the price of its products by 10 percent. But suppose the demand for the product falls by more than 10 percent. In that case, we're talking about a negative link between the product price and the company's profit. However, if the demand falls by less than 10 percent following the price rise, the link between the two elements would be positive. It's hard to predict what type of link we'll end up with in this case. This is why we call them ambiguous links.

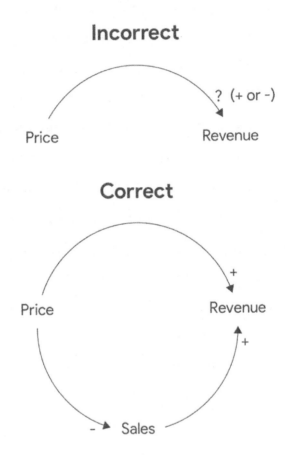

Picture 12: Ambiguous links.[xviii]

Whenever you encounter an ambiguous link, it usually means that there is more than one cause–

effect dynamic between the two elements. You need to illustrate each causal pathway explicitly in your diagram. In our case, the company's profit is impacted by the products' prices in these two ways:

- We should determine how much profit is made whenever we sell one product. (This is a positive link.)
- The price also influences how many products we sell. (This is usually a negative link. The more expensive the product is, the fewer we sell.)

We saw that when we're trying to determine whether a loop is reinforcing or balancing, it's important to be accurate with polarity labeling. Ambiguous links can give us some headaches.

Exercise

Take a look at Picture 12. What kind of loop is that, reinforcing or balancing? Why? Answer this question with both of the methods you learned.

Causation and Correlation

The function of a causal loop, as its name suggests, is to present causation within a system. Each element and each link must have some kind of causal relationship between them. One element has to have a direct effect on another to be considered a causal relationship. For example, a change in your income affects your total net worth.

Correlations need to be avoided in the causal loop diagram as they reflect past behavior, not the current system dynamics. Causal loop diagrams have to incorporate only causal relationships.

Let me give you an example to clarify what I mean. A study conducted in 2009 observed a positive correlation between the number of ice cream cones sold and the murder rate. In 2006 a similar survey connected ice cream consumption and the drowning rate. Indeed, if the number of deaths is predictably rising when people buy more ice cream cones, there must be a connection.

Would you add any of these observations to your causal loop? If you did, that would mean that, if somehow we brought ice cream selling/buying to zero, murder rates would also decrease—even to zero. While in theory it would be wonderful if things worked like that, in practice, this cause–effect relationship sounds more than ridiculous. Nevertheless, now you can understand how terribly misleading it can be when someone confuses correlation with causation. Just because one factor seemingly influences the other, it doesn't mean that is actually true. Potential correlations often are just mere coincidences.[xix]

So what on earth could be the commonality between ice cream cones and murders? The heat, for example. When it is hot outside, people buy ice cream to have a refreshing guilty pleasure. People also tend to go out more when it's warm. Therefore, they tend to clash more, and in some irreconcilable disputes, to murder more. Heat explains the increased rate of drowning, as well. Many escape to the beach when the weather is good. Would the banning of ice cream change any of the other two factors? Absolutely not.

While this correlation–causation conundrum was an easy one to figure out, most correlations are more subtle. Determining whether we talk about a correlation or causation is not always straightforward. A lot of research goes into the dissection of causation from correlation regarding the great questions of humanity. What leads to lower birth rates in developed countries: economic stability, higher child-rearing costs, education, or working women?

As you can see, correlation and causation are not always obvious. Regardless, we need to do our best to distinguish them.

Causal Loop Case Study: Menu Development

The chef of a famous restaurant is appointed by the owner to create and release a brand-new vegan menu for the chain's newest restaurant. He has to combine specific ingredients fitting today's increased demands for vegan food and match the restaurant's top-notch quality standards. The chef planned to experiment with some dishes first, and based on experts' feedback, he'd finalize the menu. He was using a small crew of cooks, dieticians, and tasters to help him.

They began working on the project and scheduled the release of the new menu for four months later. When the deadline was getting close, the chef had to delay the release until month seven as a few recipes didn't hit the restaurant's standards. In month seven, the grand opening was pushed back again to add some more dishes. In month ten, the chef rescheduled again.

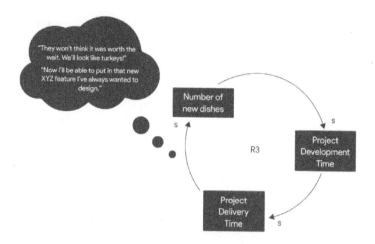

Picture 13: Menu delivery delay.

To capture this story, let's draw a causal loop diagram. Its elements are the number of new dishes, the project's development time, and its delivery date. It looks like the more the delivery date gets pushed back, the more recipes the chef wants to add to the menu. Why is this happening?

Let's dig deeper into the cognitive process of the clearly human-made decision to create more recipes and delay the opening date. We need to position the thought bubble on the line between the number of recipes and the project's delivery time.

One likely reason for the compulsive menu expansion is a cognitive bias called the sunk cost fallacy. The restaurant's owner thinks that so much time has passed and so many resources have already been invested in the vegan menu project that now he truly needs to make sure the menu is impressive. "If my menu is not above and beyond, everyone will think that the wait was for nothing."

From the chef's perspective, the delay could be an opportunity to experiment with new dishes. "Luckily, I have some extra time to try out the recipe I always wanted."

By thinking deeply enough about the hidden reasons of the actors who create a problem in the system, we can discover additional elements to add to our mental model. We could assume that the longer the delay, the more people can taste-test the sample menu. The chef can make a more informed decision about what people like and dislike in general—the more opportunities to improve the recipes on the menu, the better. The kitchen personnel can optimize storage based on the demand for certain dishes. Going deeper and deeper into the actors' motivations in the system may reveal interconnections we didn't think about before.

It is important to deeper understand the underlying thought processes that drive some decisions before trying to fix complex problems. Thanks to this understanding, we will be able to come up with a more effective intervention plan.

Rigorously examining our mental models can lead us to better insights and actions that we wouldn't have known without the digging.[xx]

Make Your Own Causal Loop[xxi]

Creating a causal loop is like telling a story. If you get a good story, you'll gain more insight into the characters' motivations, realities, and mindsets. In the following paragraphs, I'll take you through the creation of causal loops.

1. **Choose the problem.** To get started with your causal loop diagram, you need to find an exact problem in a system you want to investigate. It always sounds more interesting to examine problems that nobody seems to be able to fix. After finding the central problem, we need to collect information about its elements, interconnections, and boundaries.

For example, in one hospital, senior ER doctors asked a group of residents to improve their knowledge about wilderness medicine. The residents formed a wilderness medicine research team. The research team had to read through long documents, previous data on patient handling, and areas that lack proper resources. They had to try to optimize the existing plans and devise faster and better interventions for those injured in the wilderness.

In principle, the residents understood the long-term benefits of investing energy and time in this research. But the numerous daily tasks of their regular ER duties often kept them away from the research project. Over time, the team grew comfortable with either not preparing for their presentations or investing only a little time in superficial research. Attendance at wilderness medicine meetings started to decline.

2. **Select key elements and "scrub."** Once you have the initial story of the problem, it's time to find the most critical elements. These could be identified easily if you prepared a cluster and interconnected circle map. Whichever elements ended up having the most links could be considered key elements.

In our example, some of the most important elements could be the following: "Senior Doctors' Expectations," "Wilderness Medicine Improvement's Importance," or "Regular ER Duty Overload."

It can be helpful to interview the subjects involved in your research whenever you have the possibility. Ask about their experience, how they feel about the problem you're investigating. When you analyze the data gathered from the participants, be vigilant for expressions of subjective experience such as "too much work," "lack of time or energy," "bad management," or "having no life." Peel off the layer of subjectivity by scrubbing the expressions of qualifying adjectives. Make sure to look at your data avoiding quick judgment or premature conclusion.

3. Describe the relationship between elements.

Step three requires you to find other variables that are closely connected to every significant variable. The more interconnections you identify, the better. Look for causes, consequences, and constraints that bind these variables.

We can connect "Too Much Work" with the "Lack of Time and Energy" variable with an S line. This means they change in the same direction—the more ER workload the residents have, the more they will lack the time and energy to do their research work.

These variable pairs will be the cornerstones of our causal loop diagram.

4. **Tell the story.** We need to find several variable pairs to find a shared story running through them all. The shared story will guide us in developing a model that includes only the problem's relevant aspects. The story should have clear feedback mechanisms, which means the events presented should be recurring.

The hospital story's title could be "Improvement without Time and Energy" to portray the resident team's dynamics. The research effort failed because the people who were responsible for it were overworked and tired. For the residents, the researching task was secondary to their more urgent daily demands.

5. **Build a causal loop.** Once upon a time, a causal loop designer had to start connecting the relevant variables in the shape of a loop to create a causal loop diagram. The storyteller had to be vigilant to ask questions like "Why did this occur?" and "What did this affect?" to identify more causal connections.

In other words, create the loop diagram by adding all the variables and linking them with *s* (same direction) or *o* (opposite direction) connections. Add variables until the full story is told. Pay attention to delays where relevant.

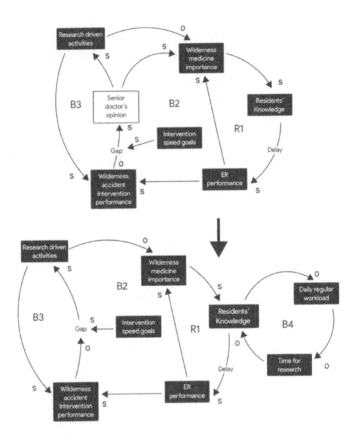

Picture 14: The system the residents work in at the hospital.

The first causal loop diagram (top) describes senior doctors' beliefs about the importance of wilderness medicine. Research and development about it would increase wilderness accident-related intervention quality and speed (R1).

However, they were also focused on research-driven activities that would reduce the gap between wilderness accident intervention performance and the overall ER performance (B2 and B3).

In the bottom diagram, Loop B4 is added to capture the dynamics of regular work demands that take the residents' time away from research activities.

Senior doctors believed that as awareness and skillfulness of wilderness accident intervention increased, the ER doctors' quality of problem handling would also increase. This would eventually lead to improved long-term medical performance, further proving the importance of research and development in specific ER fields (Loop R1). Senior doctors also wanted to make sure general ER duties were fulfilled. As the demand for research-driven activities increased, it would have taken time away from regular ER duties. This is where the residents chose to dedicate their full attention to the latter.

The link to the variable "Work Backlog" is critical. It illustrates the pressure and overwhelm that accumulates when the ER gets very busy. This balancing loop captures the residents' dynamics being regularly interrupted in their research by an emergency case in the ER. It also broadcasts the stress the residents feel for not finishing their research because they need to do their regular ER tasks (B4).

6. **Finalize the causal loop diagram.** When you're done sketching your diagram, double-check it to make sure it illustrates the story correctly. Track down logical fallacies if there are any. Make sure that all the cause-and-effect relationships are well presented.

I purposefully added "Senior Management Attention" to the loop to make this point now: it's an unnecessary variable, as it's not relevant. After the task was dispensed, the older doctors were not taking an active role in the story.

If you make any meaningful changes after revising your loop's details, go through the entire story once more. Double-check that the information gathered from the interviews is still well represented.

When designing a causal loop diagram, we focus on the accurate illustration of an individual's beliefs instead of capturing the world as it is. Systems mapping helps us gain new insights by organizing experiences and motivations in a relational framework.

Exercise

It's your turn! Try creating a causal loop about some aspect of your immediate surroundings, such as the cash flow of your bank accounts, your monthly grocery needs, trash accumulation, or even your stress levels! Add in links, labels, arrows, balancing (B), and reinforcing (R) marks as needed.

Chapter 6: Behavior-Over-Time Graphs[xxiixxiii]

Behavior-over-time graphs or BOTGs illustrate patterns and trends within a system. They show how something changes over an extended time. It is a simple tool to understand the complex nature of systems. The BOTG looks as follows:

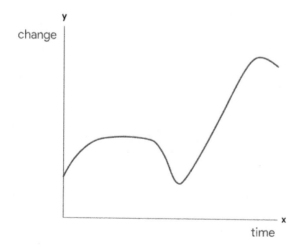

Picture 15: Behavior-over-time Graph.

The x-axis represents the passing of time, and the y-axis shows a scale of the ongoing change in the system. The line crossing the graph presents the story of the element we're examining. BOTGs give a visual explanation of our thoughts and help us understand our thinking better and draw new conclusions. We can clearly follow how specific patterns evolve and change as time passes, which gives a broader perspective than looking at isolated snapshots of events. These graphs inspire deep conversations about why and how things are changing.

Everything changes around us, always. The stock market fluctuates, an accident happens on the highway, your laptop gets slower, and your package gets delayed. These are all events affecting our everyday lives. Some we need to address quickly, but meaningful change won't happen if we stay on the event level of thinking and acting in the long run.

With behavior-over-time graphs, we can see patterns underlying the problems we're experiencing. For example, the stock market fluctuates in the short run, but historical evidence shows that over a 10-, 15-, or 20-year investment period, most people make money on it—if they are willing to endure economic highs and lows. We can check the record of the most dangerous highway sections, meaning where accidents happen more often. We can decide to avoid that chunk of it and take city roads instead. We can see the track record of our laptop brand's speed retention over time. We might discover that the laptop's longevity is not the best, so we can buy something more lasting next time. We can also switch carriers if the one we use often has delays.

We can make meaningful changes in our lives, but we need to step out of isolated event analysis and look at a longer course of patterns. BOTG graphs can be our best allies at that.

BOTG Bootcamp

Daniel Kim's following points can be used as guidelines when you're thinking of creating a BOTG.

- "**Select time horizon.** Identify the desired time horizon for the problem at hand. The length of time will provide a guide for determining which variables to select and study further. Having a time horizon of two years, for example, will have different critical variables than those associated with a time horizon of 20 years.

- **Define the problem dynamically.** Draw behavior-over-time charts of key variables. These charts can serve as reference points throughout the theory-building process, defining the problem, focusing on the conceptualization, and validating the emerging theory.

- **Conduct thought experiments.** Conduct thought experiments by hypothesizing about the time behavior of different variables and inferring other related variables' behavior. Do 'what-if' experiments of possible future scenarios and draw out those events' implications on other variables.

- **Build causal theories.** Use causal loop diagrams to build causal theories that draw out the interrelated behavior of variables over time.

- **Validate with data.** Use data analysis tools to help validate the BOTs and causal relationships."[xxiv]

Graphs are meant to illustrate a given set of data; the best-selling product of our company in the past quarter, or the Siberian tiger population of the past 50 years. When you're thinking in systems, you look at this data and ask yourself the question, "What patterns and trends can I detect?"

Was our company's best-selling product always well-performing, or has something happened in the past quarter that made it more popular? Will the Siberian tiger population grow, decline, or stay the same in the next 50 years? Why?

The best part of BOTGs is that they can provide insight regarding soft variables, as well. Soft variables are aspects of a situation that are hard or impossible to quantify. Motivation, for example, is a soft variable. Or respect. Or loyalty. While they might not seem as important to some as hard facts and numbers, it can be very impactful if these soft variables shift over time. For example, an influential manager's loyalty goes south for the company and consequently sells important information to competitors.

The BOTGs of soft variables usually tell the tale of an individual's take on the system. For example, how does the CFO feel about employees' morale in the current year when their salaries have been cut due to the pandemic?

Interrelated Patterns of Behavior—The story of …

A self-published author is confused by his sales results from the last two quarters. He was consistent with publishing his books as planned. He even finished some sooner than expected. His launch team was on fire, ready to jump into a new book project. While his book sales were not significantly declining, his profits have dropped, the first time ever since he started working as a writer. His return on investment (into promotion tools, designers, and editors) reached a record low.

He was dictating himself a fast, aggressive phase at work, so he wasn't paying too much attention to his income. Until his profits went south significantly, the author didn't realize that he had financial problems.

To address this change in his profits, the author collected data about his books' past sales performance. He looked back at a more extended time period to see if he could identify meaningful patterns. What was the sales pattern of each book launch in the past three years? How many new products had he released every four months, on average? Did he change important assistants like editors, promoters, etc.? What changes have happened in the publishing business in the past three years?

These are the answers he found: His income originating from new book launches had risen every time for the past six years, except for the last two quarters. His profit had gradually decreased in the past eight months. The self-publishing business's success was based on continuously released new products. The author kept his commitment to pushing out at least one new book each month (his quarterly average was four). He also realized that as his business grew, he hired new people to take some burden off of his shoulders. In the beginning, he ran this business 100 percent. Now he had a person in charge of promotion, one in charge of ads, a virtual assistant for emails, a social media specialist, an editor, and so on. Our author is also aware that the publishing industry is everchanging and fickle. He is selling on big platforms like Amazon and iTunes. These sites are notoriously famous for making changes in their algorithm overnight, leaving some sellers perplexed about their changed sales numbers.

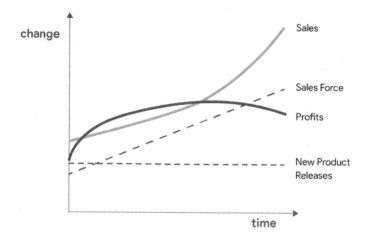

Picture 16: The author's performance over time.

Picture 16 presents the brainstorming of our author. He answered his own questions, but he didn't get to any meaningful conclusion.

Luckily, he still has a lot of uncharted space left in his BOTG. We know so far that the company was profitable for six years, minus the most recent eight months. We know that his sales dropped only relatively. Each time he publishes a new book, his overall sales rise, but in the past, each new book release amounted to 100 extra sales, now only to 70. This is still 70 more than what he'd sell without this book, but there is a tendency of decline there. That's problem number one.

Problem number two is that his profit is declining. He hired a lot of new people in the past couple of years. He keeps the phase of book launches, which means that he has to pay each of his employees for their work. But while he had 100 sales' worth of income to cover the costs in the past, now he has only 70. This thought thread would easily offer an explanation of why his profits are declining. It would be quantitative information. Let's assume that to cover the cost of launching a new book is 80 sales. Before, our author could cover all his costs and still make 20 sales' worth of profit. Now, he is losing money on each new release. This is simple math.

While we understand the numerical explanation of his profit loss, we still don't know why his sales dropped from 100 to 70. That's the question we need to find out in order to intervene and fix this problem appropriately. Firing people to reduce production costs would undoubtedly be a short-term fix. But what if the sales-dropping trend went on? What if in the next quarter our author sold only 50 books per new launch? Then 30? What if he ran out of people to fire? Then what?

Thinking deeply about this case, we can quickly conclude that firing people would rather treat a symptom than be an actual solution. The author may end up firing some staff, but whatever information we have so far is not sufficient to justify it.

This is the risk of using data analyzing tools in general … when we stick strictly to quantifiable variables, we will limit our analysis to them. Quantitative analysis tools work best when combined with some qualitative ones in the theory-building process.

Theories that combine quantitative and qualitative information are less limited by the available data and offer a more holistic picture of the problem. BOTGs are excellent tools for this purpose, as they are not data-bound. Using our BOTG above, let's try to gain some insight into the causal connections underlying the problem of this author in a two-year time frame:

- One feasible theory to understand why profits are falling is that the quality of the author's new books and their public interest have dropped since he hired new people. While the author himself took care of every aspect of his business, he might have done a better job. Maybe his employees are not the best fit for their position. Due to the pressure to produce, the author himself is spreading thin. His ideas are not so catchy anymore. There is a relationship between the number of new products and the quality of them.

- It could be that the author over-exhausted the market. Maybe the frequency of publishing should drop. He shouldn't need to pay production costs so often, and his readership might come back.

- Another possible explanation, assuming there is nothing wrong on the author's side, could be people's changed economic potential. COVID-19 hit more or less around the same time the author experienced a decline in his profits. Due to the pandemic, many were left jobless or lost most of their earning potential. That could explain a 30 percent drop in sales.

- A third theory could be a change in the algorithms. But this is such an insider aspect of big multinational companies that our author can make guesses at best.

All these theories need to be tested one at a time, of course, to know which one is accurate. Collecting the data that would support each hypothesis, we can start creating causal loop diagrams and BOTGs about each case. Testing these theories takes time, and we might jump back and forth between data analysis and theory building. This process can eventually lead us to have a better understanding of the case.

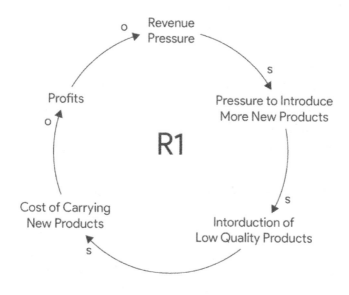

Picture 17: Causal loop of worsened quality leading to decreasing profits.

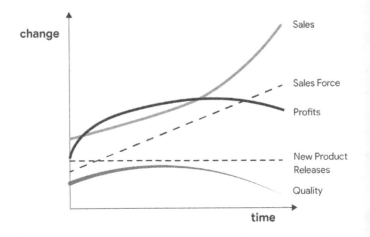

Picture 18: Adjusted BOTG showing our hypothesis about the decreased product quality.

The causal loop and BOTG above are bound to present what's happening in the author's business, considering our first hypothesis that the new books' quality has decreased.

Our theory sounds as follows, "One feasible theory to understand why profits are falling is that the quality of the author's new books and thus their public interest has dropped since he hired new people. While the author himself took care of every aspect of his business, he might have done a better job. Maybe his employees are not the best fit for their position. Due to the pressure to produce, the author himself is spreading thin, and his ideas are not so catching anymore. There is a relationship between the number of new products and the quality of them."

We can start testing this theory by changing only one variable at a time. First, the author needs to find out which of the two propositions—if either—is the real reason behind his sales drop.

1. Testing proposition A: the book quality dropped because the employees are not knowledgeable enough. He could offer courses

and training, re-explain the tasks, possible pitfalls, areas that require extra attention. The author needs to choose a set amount of time for this testing. Let's say, two months. If the numbers are not improving, three reasons could be at fault:

- This was not the real cause for the drop in sales.
- The employees are still not able to understand what they need to do.
- Two months is not enough time for the change to be visible.

Where to go from here? The author could closely monitor the employees' performance to see if they got any better and decide to replace some of them and start a new testing period. He could choose to test a totally different theory or give it more time for the change to show.

2. Testing proposition B: the author's own ideas are not the best; he's spread too thin. To fix this problem, he could interspace his new book releases, so he has more time to work on each book. He can experiment with releasing only one book every two months and see where the numbers go. If sales don't go up, it could mean three things:

- His books' quality did not improve or appeal to the audience.
- This was not the real cause for the drop in sales.
- The time he allowed for testing was not enough for the change to be visible.

What can the author do now? He could involve a bunch of free beta readers to give objective feedback on his work. He could choose to test a totally different theory or give it more time for the change to show.

Exercise

Using what you learned here about BOTGs and causal loops, try to illustrate the other three hypotheses. How would each of them be illustrated? How would you test them? What interventions would you propose to mitigate their effect?

For reference, here are the other three theories:

1. It could be that the author over-exhausted the market. Maybe the frequency of publishing should drop. He shouldn't need to pay production costs so often, and his readership might come back.

2. Another possible explanation, assuming there is nothing wrong on the author's side, could be

people's changed economic potential. COVID-19 hit more or less around the same time the author experienced a decline in his profits. Due to the pandemic, many were left jobless or lost most of their earning potential. That could explain a 30 percent drop in sales.

3. A third theory could be a change in the algorithms. But this is such an insider aspect of big multinational companies that our author can make guesses at best.

We don't want to encounter the same problems over and over again. To step out of this process, we need to see the bigger picture in which these problems exist. Behavior-over-time graphs and corresponding causal loop diagrams can help us see what has happened, and why. We can use them to create predictions for the future and intervene in the system accordingly.

Chapter 7: Stock-and-Flow Diagrams

Let's recapitulate what we know about systems so far. Systems are comprised of elements (or variables) that are interconnected and work towards achieving an outcome. We call this the purpose or function of the system. We also know that the events within the system are dynamic. They arise because of a specific cause or causes; we can trace this with causal loop diagrams. These events influence the system's behavior, which over time can reveal some patterns. We can analyze this with behavior-over-time graphs. Thanks to the repeating patterns, we can decide if the system is affected by reinforcing or balancing feedback mechanisms.

All that I have described so far is a part of what we call the dynamic systems model. Stocks and flows are essential parts of this model.

A stock is an element (or variable) that is measured at one given moment in time. It shows the quantity of the element present at that point in time.

A flow is measured over an interval of time. Thus, we would attribute units of time measurement to our flows, for example, over the last month or year.[xxv]

Here is another way of explaining stock and flows. Donella Meadows, in her book, *Thinking in Systems: A Primer*, talks about stocks and flows as being the foundation of every system.[xxvi]

Stock

A stock can be something physical, like the amount of money in your account, your business inventory, or the information needed to write a book. However, stocks don't necessarily have to be physical things. They can also be feelings or attitudes people hold. The amount of anger you hold toward a cheating spouse or an alcoholic parent can also be viewed as a stock.

Stocks are not static. They change over time based on the impacts of a flow. Stocks are snapshots in time, showing the current condition of the changing flows in the system.

Flow

Flows are the actions that impact a system. A flow might be a success or a failure, purchases or sales, deposits or withdrawals, births or deaths, or growths or declines. To illustrate how flows impact the stocks, I've added Picture 19 below. The rectangle stands for the stock. This stock is affected by two flows, specifically an inflow and an outflow.

Stocks can be affected by multiple inflows and outflows, but for the sake of simplicity, let's stick to only one of each. The little clouds represent where the flows come from or go. Flows can be increased or decreased through intervention. This is what the arrow labeled feedback is deemed to represent. How does this look like? Let's take a look at the symbols first.

Name	Symbol
Stock	

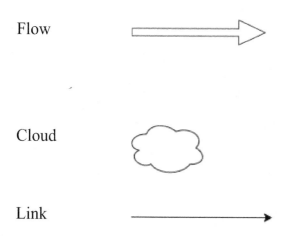

Flow

Cloud

Link

The cloud is either meant to stand for the source of a flow (when the flow is coming from outside sources), or it shows the sink of the flow.

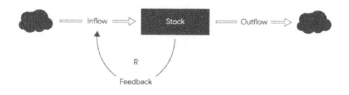

Picture 19: Sample stock-and-flow diagram.[xxvii]

Let's look at a real case study.

Picture 20 (below) shows Ian's money in an interest-bearing bank account (stock). The inflow is his paycheck, and the outflow is his monthly spending. By choosing to put his stock (your savings) into an interest-bearing account, the stock will affect the inflow by reinforcing feedback. The interest he earns will become a part of his income, in other words. What does this mean? It means that the bigger Ian's stock (the more money he has in the interest-bearing account), the more inflow (salary plus money earned through interest) he'll have. The more inflow he has, the bigger his stock grows. Our reinforcing feedback serves as a tool to better understand the dynamics of the stock and flows.

Picture 20: The relationship between stock and flows in action.[xxviii]

Some simple conclusions we can draw from this diagram:

- If you earn more than you spend, your account will grow.

- If you spend more than you earn, your bank account shrinks. If this situation persists, you'll end up in debt.

- If you make $100 and spend $100, you'll have the same amount of money in your account in April as you had in March.

- If you cut your spending and find an additional revenue source, your bank account will grow even faster.[xxix]

Let's take our knowledge of stocks and flows further. As I mentioned before, there are cases when there are multiple (in and out) flows affecting the stock, but there are also cases when a stock has only one flow. Take nonrenewable resources, like a newly discovered oil field, as an illustrative example. As oil's natural creation takes hundreds of thousands of years, it is safe to say the recently discovered oil field (stock) will only have outflows. The outflow, in this case, is oil extraction. The quicker we extract the oil, the faster we deplete the field.

Now let's imagine the opposite. Think of a comic book collector who would rather die than give away any of their precious items. Assuming the paper doesn't deteriorate over time, and the collector is a vampire who lives forever, the comic book collection stock doesn't have an outflow, only an inflow. This means our Nosferatu will only add new items to their collection. This could be a hypothetical example of a stock that only has inflows. In real life, however, such cases hardly ever exist—or persist.

The general rules of stocks and flows:

- If there are more inflows than outflows, the level of stock will increase.
- If there are more outflows than inflows, the level of stock will decrease.
- If the number of outflows and inflows is equal, the stock level will remain at its current level and will be unchanged. This is called dynamic equilibrium.

- The level of a stock is increased if its outflow is decreased or its inflow is increased.

- Stocks provide a security barrier in a system since they delay the initial shock that may affect a system.

- Stocks preserve the ability of inflows and outflows to remain independent.

Our mind has a tendency to focus more on stocks than the flows. Stocks are more tangible and visible than the flows that affect them. Even when we look at the flows, we are more likely to focus on the inflows than outflows.

This could lead to forgetting that there is more than one way to get our stock to the desired level.

What do I mean by that?

Let's keep the bank account as our example. We are more prone to wish to grow our fortune by increasing the money inflow. It is much more difficult and unnatural to acknowledge that decreasing our outflow, spending less, also helps us reach our goal. Sure enough, there is only a certain amount of restriction we can put on our money outflow. At the same time, the inflow can be hypothetically limitless.

We can change flows quickly if we want, but stocks react more slowly to change. We can eat a piece of chocolate (inflow) and then go for a half-hour jog (outflow) to get rid of the extra calories. But our weight (stock) doesn't instantly drop or rise. We can plant one hundred trees in a short period, but it will take decades for those trees to grow into a forest. Areas affected by droughts do not immediately see their reservoirs return to their normal water levels. Nor are the negative impacts of global warming instantly reversed.

By their slowly changing nature, stocks act as buffers, lags, or delays in the system—the bigger the stock, the slower the change. People often don't take the nature of stocks into consideration. It takes time to build highway systems, to improve the infrastructure, or to boost the economy.

Stock changes set the pace for system dynamics. A highway won't be ready quicker than the workers' speed, the building materials' solidification, and of course, all the bureaucratic processes.

From a system mapping point of view, it is imperative to understand how quickly a stock changes to the desired level. Suppose we want to save up a million dollars. In that case, that won't happen overnight (unless we win the lottery or we get lucky at blackjack in Vegas). If we know what income we can expect each month, how much we can save, and how compound interest works in our favor, based on this data, we can calculate how many years it will take our stock to turn into one million dollars.

Other examples of stock accumulation could be the number of calories we eat each day, the number of steps we take, or the amount of water we fill our bathtub with. Interestingly, more abstract concepts can act as stock, too. And they can accumulate or shrink—for example, our stress level, our relationship satisfaction, or our goodwill towards someone.

Stock-and-flow system maps help us notice the accumulations, drops, and the speed at which their rate changes. We can use these maps to analyze questions in our own lives, like the bank account example. But in the bigger scheme, system dynamics experts use these diagrams in a mathematical computer program to represent a system of interest. They can map out anything from world-level poverty fluctuation to the speed of spread of a virus, for instance. Stock-and-flow diagrams provide visual illustrations that help us explain the causal connections contributing to the analyzed system's behavior and lead us to find potential intervention areas.

How to Convert Causal Loop Diagrams into Stock-and-Flow Diagrams[xxx]

Your first question might be, why should we convert causal loop diagrams into stock-and-flow diagrams? Well, thank you for asking. There is a strong connection between these two. In the following pages, I answer this question while I guide you through the transformation process itself. This is an important step in strengthening our knowledge about stocks and flows, highlighting the difference between information and material flows. We will also see why unit consistency is imperative throughout our diagram.[xxxi]

Step 1: Specify the Units of Every Element in your Causal Loop Diagram

Let's stick to the example of a self-published author who is selling books. What are the elements in this system? What would be the units of these elements?

How can we measure something? We need to know the unit of what we're measuring. It's critical to know the unit of the elements because:

- It helps us think about the causal loop in a more specific manner, which helps translate the loop into a stock-and-flow diagram later.

- We can assess which elements are related to time—these will probably be flows. We will also notice more easily which elements are missing and need to be added later to the stock-and-flow diagram.

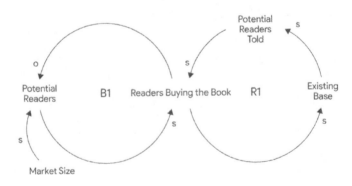

Picture 21: Causal loop diagram for a book's life cycle.

In Picture 21 we can see the causal loop diagram for a book's life cycle. The unit for "Installed Base," "Potential Customers," and "Market Size" is people. The unit for "People Buying Product" is people per month.

Step 2: Identify the Stocks

Once we've established the units of the elements, it's time to find our stocks. As you can see, the previous step helped us a lot in that. We know that the time-related elements are likely flows. (Be cautious. You need to check the causal loop diagram's overall function before concluding if the element is indeed a flow.)

What we determined to be a unit of people are likely our stocks. Go through the thought process of the causal loops and determine if any additional stocks are needed. In the book's life-span diagram, we can separate two stocks, the "Potential Customers" and the "Installed Base."

Step 3: Identify the Flows

Which elements are the ones that add or take away from the stocks? Only flows can do that. If one element in the causal loop diagram affects the size of a stock over time, it's a good guess that we found our flow. In the example above, this would be "People Buying Product."

Step 4: Connect What You Can

Let's break down this step. First, we need to link the flow to every stock it affects. How can we know if the flow is an outflow or inflow? If the stock is negatively influenced by the flow, it is an outflow; if the stock is positively affected, we're facing an inflow. Looking at our example, when people buy more books (the flow of "People Buying Product" increases), the author's readership ("Installed Base" stock) also increases. Therefore "People Buying Product" is an inflow to "Installed Base." (See Picture 22.)

People Buying the Book   Existing Base

Picture 22: Inflow.

Simultaneously, as "Installed Base" increases, the "Potential Customers" stock decreases as some of them became actual readers. So, we're facing an outflow.

Potential Readers Installed
People Buying the Book
Existing Base

Picture 23: Outflow and inflow.

When we're done connecting our flows to stocks, it may be necessary to connect some stocks to flows. We need to do this when a stock affects one or more flows. We link them. In a causal loop diagram, the same type of link can be related to either carrying physical things (e.g., the paper for print books) or information (for example, book quality). A stock-and-flow diagram calls for separating such kind of links. In our case, there are no information links that connect the stocks to the "People Buying Product" flow.

Step 5: Add and Link the Remaining Elements of the Causal Loop Diagram

As you could see, after identifying the stocks and flows, there were some leftover elements in the causal diagram loop, like "Market Size."

We call them "auxiliary" elements. They come in two versions:

1. **Constants:** elements whose values won't change during the time interval we're analyzing.

2. **Elements originating from stock-and-flow-based calculations.** In the case of the book's life-span diagram, "Market Size" is an auxiliary element, and it is constant (we assume this to keep our model simple). Another auxiliary element could be "Percentage of Market Untapped"—this would be the result of "Potential Customers" divided by "Market Size."

Link the new elements to the elements that they affect and to those that they are influenced by. Stocks can influence auxiliary elements; "Potential Customers" affects "Percentage of Market Untapped." It doesn't work the other way around—auxiliary variables can't affect stocks. Only flows can. If you are confident that the auxiliary variable affects the stock, it may be a flow in disguise.

By now, your stock-and-flow diagram looks like this:

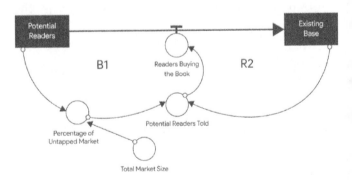

Picture 24: Initial stock-and-flow diagram of the book's life-span.

This is not our final version. We will need to examine the interrelationships further before deeming the diagram complete. Something important is missing. The final goal of a stock-and-flow diagram is to show information in a calculable fashion. One of the key differences between a causal loop diagram and a stock-and-flow diagram is that we can see how the elements are linked in the former. We can't yet calculate the value of one element, given the value of the other elements. In the latter, however, we get a calculable representation of the system.

Step 6: Defining Elements, Checking the Units

When defining elements, our first move should be to specify the equation that helps us measure the elements' value. To do this, we need to know the elements' initial value and the value of every other element present in the diagram.

Stocks are the easiest to define. The calculation looks like this: we add the effects of the outflows and inflows to the amount already in the stock. So if we have the initial stock level and the unit in which we make the calculation, we're good to go. How do we find the unit, again? In our "book's life span" example, if we measure "Installed Base" in people and "People Buying Product" in dollars, then the unit of the latter is incorrect.

After the stocks are defined, we need to do the same check with the flows. In our case, the flow "People Buying Product" is defined as the number of people buying the book each month. Thus the unit is people per month. This flows into the stock "Installed Base." This stock represents the full amount of people who bought the book.

Step 7: Additional Variables

Now we know how our stocks and flows are defined. We know what units and measurements they come in. The only thing left to do is analyzing the leftover elements. We need to define them and check their measurement units too to make sure our diagram is consistent. Like in step 6, go through the auxiliary elements and see if they come in the same unit as the main stock and flow. You may discover additional elements in this process. If you do, add them to the diagram.

Once everything is defined and unit-consistent, congratulations! You converted a causal loop diagram into a stock-and-flow diagram! One small thing before you celebrate prematurely, the model is still not calculable.

How would you calculate "People Buying Product"? What number of "Potential Customers" actually buy the book on average per month?

If we don't know this number, we can't measure our flow's value, "People Buying Product." An element is missing here. What is missing is the percentage of people told about the book who end up buying it.

We can name this element "Probability of Potential Readers to Buy." With the help of this element, if we know how many potential readers have been told about the book ("Potential

Customers"), we can calculate "People Buying Product."

We're not done yet! It turns out we still can't calculate the "Potential Customers" element as we don't know how many people each existing reader ("Installed Base") referred our book to. Without that data, how could we figure out the number of potential customers told about our book?

You are correct. We need another auxiliary element. Let's call it "Referral per Reader." If we get this number, we can calculate "Potential Customers" (the sum of referrals made by each person in our "Installed Base" multiplied by "Percentage of Market Untapped"). We can then use that number and "Probability of Potential Readers to Buy" to calculate "People Buying Product."[xxxii]

Now we have a calculable diagram.

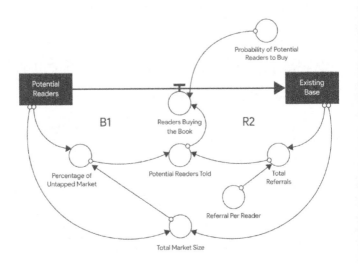

Picture 25: The final version of "book's life span."

Stock-and-flow diagrams help you see the subtleties of the system. These diagrams work with the most important elements and need you to find specific information about them, such as their units and relative magnitude. Stock-and-flow diagrams also make you brainstorm more about the interconnections between elements and see if their units mix properly. You will also be somewhat forced to find auxiliary elements that are needed to make the units match. All these steps give you a more detailed diagram and a better understanding of the system.

In Closing

In this book, we took a closer look at five systems thinking tools:

- cluster maps
- interconnected circle maps
- causal loop diagrams
- behavior-over-time graphs
- stock-and-flow diagrams

Each of these is a very helpful standalone tool. But build them on one another and use them in combination. You will gain deeper insights into dynamic behavior.

I wish you much success in your practicing.

A.R.

Visit www.albertrutherford.com and download your FREE copy of the booklet, The Art of Asking Powerful Questions in the World of Systems!

Reference

Acaroglu, L. (2017, September 20). *Tools for Systems Thinkers: Systems Mapping.* Medium. https://medium.com/disruptive-design/tools-for-systems-thinkers-systems-mapping-2db5cf30ab3a.

Aronson, D., & Angelakis, D. (n.d.). *Step-By-Step Stocks and Flows: Converting From Causal Loop Diagrams.* https://thesystemsthinker.com/step-by-step-stocks-and-flows-converting-from-causal-loop-diagrams/.

Forrester, J. W. (1989, July 13). *The Beginning of System Dynamics* [Banquet Talk]. International Meeting of the system Dynamics Society, Stuttgart, Germany. https://web.mit.edu/sysdyn/sd-intro/D-4165-1.pdf.

Foundations of Systems Thinking. (n.d.). *Tool #1: Behavior-Over-Time Graphs.* Foundations of Systems Thinking. https://foundationsofsystemsthinking.org/courses/tools/01-behavior-over-time-graphs/making-thinking-visible/.

Foundations of Systems Thinking. (n.d.). *Tool #3: Stock-Flow Mapping.* https://foundationsofsystemsthinking.org/cour ses/tools/03-stock-flow/accumulation/.

Frangos, A. (2013, June). *Think Like a Modeler Skill #2: Expand.* Systems & Us. https://systemsandus.com/2013/06/05/think-like-a-modeler-skill-2-expand/.

Ho, L. (n.d.). *Ice Cream Sales "lead" to Homicide: Why?.* Lifehack. https://www.lifehack.org/624604/the-most-common-bias-people-have-that-leads-to-wrong-decisions.

Karash, R. (n.d.). *Mental Models and Systems Thinking: Going Deeper Into Systemic Issues.* The Systems Thinker. https://thesystemsthinker.com/mental-models-and-systems-thinking-going-deeper-into-systemic-issues/.

Kim, D. (n.d.). *Behavior over Time Diagrams: Seeing Dynamic Interrelationships.* The Systems Thinker. https://thesystemsthinker.com/behavior-over-time-diagrams-seeing-dynamic-interrelationships/.

Kim, D. (n.d.). *Guidelines for Drawing Causal Loop Diagrams.* The Systems Thinker. https://thesystemsthinker.com/guidelines-for-drawing-causal-loop-diagrams-2/.

Kim, D. (n.d.). *Using Causal Loop Diagrams to make Mental Models Explicit.* The Systems Thinker. https://thesystemsthinker.com/using-causal-loop-diagrams-to-make-mental-models-explicit/.

Learn Systems Thinking. (n.d.). *An Example of a Stock and Flow Diagram.* https://learnsystemsthinking.weebly.com/diagrams.html.

Meadows, D. H. (2008). *Thinking in Systems.* Earthscan Publications.

Molloy, J. (n.d.). *Learning About Connection Circles.* The Systems Thinker. https://thesystemsthinker.com/learning-about-connection-circles/.

Sterman, J. (n.d.). *Fine-Tuning Your Causal Loop Diagrams—Part 1.* The Systems Thinker. https://thesystemsthinker.com/fine-tuning-your-causal-loop-diagrams-part-i/.

Stroh, D. P. (2006). *Defining Variables.* Applied Systems Thinking. https://www.appliedsystemsthinking.com/sup

porting_documents/Practice_DefiningVariabl es.pdf.

System. (n.d.). In *Mirriam-Webster.com dictionary*. Retrieved January 1, 2021, from https://www.merriam-webster.com/dictionary/system.

Systems & Us. (n.d.). *What is a Model and Why Learn Modeling?* https://systemsandus.com/foundations/why-you-should-think-like-a-modeler/what-is-a-model/.

Zhou, J. (2013, June). *Think Like a Modeler Skill #1: Define.* Systems & Us. https://systemsandus.com/2013/06/13/think-like-a-modeler-skill-3-define/.

Zhou, J. (2013, May). *Think Like a Modeler Skill #3: Aggregate.* Systems & Us. https://systemsandus.com/2013/05/23/think-like-a-modeler-skill-1-aggregate/.

Zulkepli, J., Eldabi, T., & Mustafee, N. (2012, December). *Hybrid simulation for modelling large systems: an example of integrated care model* [Conference Proceedings]. Winter Simulation Conference, Berlin, Germany.

Endnotes

[i] Forrester, J. W. (1989, July 13). *The Beginning of System Dynamics* [Banquet Talk]. International Meeting of the system Dynamics Society, Stuttgart, Germany. https://web.mit.edu/sysdyn/sd-intro/D-4165-1.pdf.

[ii] System. (n.d.). In *Mirriam-Webster.com dictionary*. Retrieved January 1, 2021, from https://www.merriam-webster.com/dictionary/system.

[iii] Zulkepli, J., Eldabi, T., & Mustafee, N. (2012, December). *Hybrid simulation for modelling large systems: an example of integrated care model* [Conference Proceedings]. Winter Simulation Conference, Berlin, Germany.

[iv] Stroh, D. P. (2006). *Defining Variables*. Applied Systems Thinking. https://www.appliedsystemsthinking.com/supporting_documents/Practice_DefiningVariables.pdf.

[v] https://thesystemsthinker.com/causal-loop-construction-the-basics/

[vi] Meadows, D. H. (2008). *Thinking in Systems*. Earthscan Publications.

[vii] Systems & Us. (n.d.). *What is a Model and Why Learn Modeling?* https://systemsandus.com/foundations/why-you-should-think-like-a-modeler/what-is-a-model/.

[viii] Zhou, J. (2013, June). *Think Like a Modeler Skill #1: Define.* Systems & Us. https://systemsandus.com/2013/06/13/think-like-a-modeler-skill-3-define/.

[ix] Frangos, A. (2013, June). *Think Like a Modeler Skill #2: Expand.* Systems & Us. https://systemsandus.com/2013/06/05/think-like-a-modeler-skill-2-expand/.

[x] Zhou, J. (2013, May). *Think Like a Modeler Skill #3: Aggregate.* Systems & Us. https://systemsandus.com/2013/05/23/think-like-a-modeler-skill-1-aggregate/.

[xi] Zhou, J. (2013, May). *Think Like a Modeler Skill #3: Aggregate.* Systems & Us. https://systemsandus.com/2013/05/23/think-like-a-modeler-skill-1-aggregate/.

[xii] Acaroglu, L. (2017, September 20). *Tools for Systems Thinkers: Systems Mapping.* Medium. https://medium.com/disruptive-design/tools-for-systems-thinkers-systems-mapping-2db5cf30ab3a.

[xiii] Picture 5. Acaroglu, L. (2017, September 20). *Tools for Systems Thinkers: Systems Mapping.* Medium.

https://medium.com/disruptive-design/tools-for-systems-thinkers-systems-mapping-2db5cf30ab3a.

xiv Molloy, J. (n.d.). *Learning About Connection Circles*. The Systems Thinker. https://thesystemsthinker.com/learning-about-connection-circles/.

xv Sterman, J. (n.d.). *Fine-Tuning Your Causal Loop Diagrams—Part 1*. The Systems Thinker. https://thesystemsthinker.com/fine-tuning-your-causal-loop-diagrams-part-i/.

xvi Kim, D. (n.d.). *Guidelines for Drawing Causal Loop Diagrams*. The Systems Thinker. https://thesystemsthinker.com/guidelines-for-drawing-causal-loop-diagrams-2/.

xvii Kim, D. (n.d.). *Guidelines for Drawing Causal Loop Diagrams*. The Systems Thinker. https://thesystemsthinker.com/guidelines-for-drawing-causal-loop-diagrams-2/.

xviii Sterman, J. (n.d.). *Fine-Tuning Your Causal Loop Diagrams—Part 1*. The Systems Thinker. https://thesystemsthinker.com/fine-tuning-your-causal-loop-diagrams-part-i/.

xix Ho, L. (n.d.). *Ice Cream Sales "lead" to Homicide: Why?*. Lifehack. https://www.lifehack.org/624604/the-most-common-bias-people-have-that-leads-to-wrong-decisions.

xx Karash, R. (n.d.). *Mental Models and Systems*

Thinking: Going Deeper Into Systemic Issues.
The Systems Thinker.
https://thesystemsthinker.com/mental-models-and-systems-thinking-going-deeper-into-systemic-issues/.

[xxi]Kim, D. (n.d.). *Using Causal Loop Diagrams to make Mental Models Explicit.* The Systems Thinker. https://thesystemsthinker.com/using-causal-loop-diagrams-to-make-mental-models-explicit/.

[xxii] Foundations of Systems Thinking. (n.d.). *Tool #1: Behavior-Over-Time Graphs.* Foundations of Systems Thinking. https://foundationsofsystemsthinking.org/cour ses/tools/01-behavior-over-time-graphs/making-thinking-visible/.

[xxiii] Kim, D. (n.d.). *Behavior over Time Diagrams: Seeing Dynamic Interrelationships.* The Systems Thinker. https://thesystemsthinker.com/behavior-over-time-diagrams-seeing-dynamic-interrelationships/.

[xxiv] Foundations of Systems Thinking. (n.d.). *Tool #1: Behavior-Over-Time Graphs.* Foundations of Systems Thinking. https://foundationsofsystemsthinking.org/cour ses/tools/01-behavior-over-time-graphs/making-thinking-visible/.

[xxv] Foundations of Systems Thinking. (n.d.). *Tool #3:*

Stock-Flow Mapping.
https://foundationsofsystemsthinking.org/cour
ses/tools/03-stock-flow/accumulation/.

xxvi Meadows, D. H. (2008). *Thinking in Systems.*
Earthscan Publications.

xxvii Picture 19. Learn Systems Thinking. (n.d.). *An
Example of a Stock and Flow Diagram.*
https://learnsystemsthinking.weebly.com/diagrams.ht
ml. Downloaded 2020.

xxviii Picture 20. Learn Systems Thinking. (n.d.). *An
Example of a Stock and Flow Diagram.*
https://learnsystemsthinking.weebly.com/diag
rams.html. Downloaded 2020.

xxix Meadows, D. H. (2008). *Thinking in Systems.*
Earthscan Publications.

xxx Aronson, D., & Angelakis, D. (n.d.). *Step-By-Step
Stocks and Flows: Converting From Causal
Loop Diagrams.*
https://thesystemsthinker.com/step-by-step-
stocks-and-flows-converting-from-causal-
loop-diagrams/.

xxxi Aronson, D., & Angelakis, D. (n.d.). *Step-By-Step
Stocks and Flows: Converting From Causal
Loop Diagrams.*
https://thesystemsthinker.com/step-by-step-
stocks-and-flows-converting-from-causal-
loop-diagrams/.

xxxii Aronson, D., & Angelakis, D. (n.d.). *Step-By-
Step Stocks and Flows: Converting From
Causal Loop Diagrams.*

https://thesystemsthinker.com/step-by-step-stocks-and-flows-converting-from-causal-loop-diagrams/.